Healed & Whole

31-DAY TRANSFORMATION

IdellaLiselle.com

Idella Liselle

@Idella_Liselle

Idella Liselle

www.linkedin.com/in/idellaliselle

Follow for Prayer & More......

idella Liselle
The CEO Coach

Published by:
Publish*her* Publishing
Idella Liselle | www.idellaliselle.com

Cover Design: PublishHer Publishing
Interior Design: PublishHer Publishing

First Edition, 2025
Printed in the United States of America

ISBN: 978-1-948731-30-0

This devotional is not a substitute for professional counseling or medical advice. It is intended to accompany personal healing, spiritual reflection, and participation with the HOLY SPIRIT on your journey to wholeness.

This Is Not a Pretty Journey

Healing is not a polished process. It's not always linear, and it's rarely Instagram-worthy. Healing is raw. It's messy. And it starts with one decision: to be real.

Real about what broke you.
Real about what you've been pretending doesn't hurt.
Real about the patterns you've normalized to survive.

This devotional is not about surface-level recovery. It's about divine surgery. It's about giving the HOLY SPIRIT full access—not just to your Sunday self, but to the silent corners, the secret wounds, the stories you don't say out loud. Healing and wholeness only come when we stop managing pain and start surrendering it.

You are not reading this by accident. Something in you—maybe a whisper, maybe a scream—said I want to be whole. But here's the truth: GOD won't heal what we won't hand over. And wholeness won't come by wishing. It comes through obedient participation. That means showing up daily, even when it hurts. It means telling the truth, even to yourself. It means letting the HOLY SPIRIT work in layers, not just symptoms.

So if you're ready—not to be perfect, but to be present—this journey is for you.

Let this be your prayer before you begin:

Father,

I don't want to play pretend. I give You full access to every wound, every memory, and every lie I've believed. I'm not here to impress—I'm here to be made whole. I invite You into the ugly, the hidden, and the forgotten places. Break through my defenses. Heal what I buried. Reveal what I need to release.

I surrender my pride, my pace, and my preferences. I declare: I am ready to participate. Speak to me. Heal me. Transform me.

I partner with HOLY SPIRIT! The One who leads and guides to all Truth. Touch the parts of me I've been afraid to face. Pull up the roots I covered with excuses. Unwind the patterns I wrapped around my pain. Don't let me confuse survival with healing. Rebuild me from the inside out—layer by layer, lie by lie, memory by memory—until nothing in me contradicts Your truth.

I want to be healed and whole.
In JESUS CHRIST name, Amen.

If you're ready to grow beyond these pages, we invite you to join our group and let's journey together with a community of others who are also choosing healing and wholeness. Join the Healed & Whole group on Telegram—where we grow, share, and heal together.

You're not alone. Let's keep going—together.

Telegram Group: Healed & Whole
Link: https://t.me/+SCNaXelEDZc5ZWMx
Or Scan this QR:

Prepare for the Process

You're not just reading a devotional — you're stepping onto a battlefield for your healing and your wholeness.

This 31-day war cry won't just show you the places that need healing — it will drag into the light every system, lie, and strategy that's tried to choke it out. The moment you say "yes" to healing, expect resistance. Internal sabotage will scream. External pressure will rise. Life will throw everything it can to interrupt your devotion. The enemy will whisper that nothing's changing, that this is too heavy, that maybe you should quit. But hear me: that's not failure — that's proof you've hit the real roots.

Healing is not cosmetic work. Wholeness is demolition. It rips through hidden agreements, digs up suppressed wounds, breaks generational shackles, and tears down emotional scaffolding that's been holding you hostage.

This is going to confront your avoidance. It's going to demand your voice, your action, your sweat. Healing is not a handout. Wholeness doesn't bow to passivity. It is activated by your agreement, your participation, and your obedience. You're not being healed to feel better — you're being healed to be better. You're not recovering crumbs — you're reclaiming the crown of who you were always designed to be.

You will legislate your healing. With the authority of GOD's Word, the partnership of HOLY SPIRIT, and the decrees that leave your mouth, you will shift atmospheres.

Wholeness is not the end — it's the beginning of a new era where cycles break, patterns end, and identity reigns. You will walk differently because the chains that shaped your steps are gone. You will see differently because shame no longer clouds your vision. You will speak differently because your voice now legislates Kingdom realities.

This isn't about survival anymore. It's about dominion. What once wounded you will now witness your authority. What once buried you will now become your building ground. This war ends with you whole — and the world changed because you are.

As you confront, you conquer.
As you surrender, you recover.
As you show up, you possess.
You are healed and whole!!

DATE:

You are about to reclaim your full capacity.
You are about to become
HEALED & WHOLE.

AREAS I WANT TO BE HEALED & WHOLE

Notes:

Heart Prep

Father, I step into this journey as a willing participant in my healing and wholeness. I refuse to approach this process casually, but instead, I posture myself in full surrender and active participation with the HOLY SPIRIT. I give You permission to search, expose, and confront every hidden place in my soul that has resisted healing. I will not avoid the deep places; I will confront what must be conquered.

I cancel every agreement I have made with denial, delay, comfort zones, or survival mechanisms that have kept me functioning beneath divine capacity. I sever ties with excuses, self-protection, emotional apathy, and spiritual stagnation. I yield to the work that brings wholeness — even when it challenges me. I embrace truth, even when it confronts my comfort. I receive correction, even when it demands change.

Father, train my eyes to see, my ears to hear, and my heart to perceive what You reveal in these 31 days. I decree that what is exposed will be healed, what is confronted will be conquered, and what is uprooted will be replaced with Your truth. I release the angels assigned to this healing journey to guard my mind, soul, and spirit as I legislate my wholeness.

I decree that I will complete this process, fully healed, fully whole, fully aligned, fully activated. Nothing in me will remain half-restored. This process will produce permanent fruit, and my wholeness will testify of Your power — in JESUS CHRIST name, Amen.

DAY ONE

Jeremiah 30:17 AMPC

For I will restore health to you and I will heal your wounds, says the LORD, because they have called you an outcast, saying, This is Zion, whom no one seeks after and for whom no one cares!

Let's Talk About It:

There are wounds that bandages can't fix. The kind of pain that gets buried under "I'm fine" and "I'm used to it." This verse isn't just about physical healing —it's about the kind of healing that finds you after you've been labeled, rejected, and forgotten. GOD says He's coming for that wound. The one nobody sees but has shaped how you walk through life. He's not intimidated by what hurt you. In fact, He's already made plans to restore what was broken when they walked away.

But let's be honest—when your name has been dragged through dirt, when people have defined you by your lowest moment, when the whispers become louder than your own thoughts—it's hard to believe restoration is possible. You start adapting to dysfunction. You settle for survival. But GOD doesn't just want to heal the injury—He wants to erase the evidence. He doesn't just restore your health—He restores your name. He calls you healed when they called you broken. He calls you chosen when they left you out. And when GOD calls you by a new name, the old one loses its power. This healing is personal. It goes to the place where you stopped believing things could change—and it rewrites the story.

Reflection Question:

What wound have you hidden so well—even from yourself—that you stopped expecting GOD to ever touch or heal it?

____

____

____

____

____

Challenge:

I challenge you to sit with the silence—the silence you've been drowning in busyness. Sit and listen to the part of you that feels unloved, unwanted, and misunderstood. Open yourself fully to the moment. Let the HOLY SPIRIT expose every lie that rejection has planted in your life, lies that have camouflaged themselves as part of your identity.

Call those lies out. Call them to death. Use the sword of the SPIRIT to cut the head off every voice that still echoes rejection. And if it's a hydra spirit—one lie feeding into another—then keep swinging until every head is severed. This is your war cry for healing.

____

____

____

____

____

PROPHETIC DECREE

I present every area of my life marked by rejection, abandonment, and mislabeling. Every wound created by being called an outcast is exposed to the healing power of GOD. I forgive and release those who wounded me. I sever their influence from my identity and uproot every experience still tied to their words.

I reject the weight of rejection and the lie that I am unwanted, unseen, or unworthy. I stand under the authority of Jeremiah 30:17 and decree that the LORD is restoring my health and healing every wound—visible and hidden. I will no longer carry affliction as a condition or a personality.

I come into agreement with the prosperity of my soul. I engage with the HOLY SPIRIT to stabilize every fractured part of me. I open myself fully to divine recovery.

The Great Physician is reversing trauma, dismantling infirmity, and rebuilding my health with precision. This reconstruction is not cosmetic—it is cellular, emotional, spiritual, and final. My health is not delicate. It is reinforced by the covenant.

By apostolic authority, I call forth the full manifestation of healing across every timeline and dimension of my life. I declare that generational rejection is terminated, emotional residue is extracted, and spiritual scar tissue is dissolved by the fire of GOD. Every label that did not come from GOD is now illegitimate and stripped of power. I renounce false names, cursed agreements, and traumatic adaptations.

I legislate alignment between my soul and GOD's original design. I release a divine summons to every dormant part of me—arise and be restored. I authorize the angelic to minister to the places man misdiagnosed and to bring reinforcement where I once bled silently.

What broke me no longer binds me. What silenced me can no longer speak. I am not rehabbing—I am being rebuilt by the WORD and through the BLOOD.

I am not just recovering—I am reigning.

I am healed and whole in JESUS CHRIST name, Amen.

Thoughts:

Thoughts:

Thoughts:

DAY TWO

Jeremiah 33:6 KJV

Behold, I will bring it health and cure, and I will cure them, and will reveal unto them the abundance of peace and truth.

Let's Talk About It:

There's a kind of dysfunction you didn't choose. It was modeled. Inherited. Normalized. You adapted to survive. And now, as you try to heal, you're facing invisible systems within your soul that resist health because they were taught to function in chaos.

But the promise still stands: "I will bring it health and cure…"

The healing GOD offers is not random—it is revelatory. He reveals peace and truth, which means the dysfunction must be exposed. And that exposure is not to shame you. It's to retrain your soul.

This is the season where the HOLY SPIRIT doesn't just address symptoms—He confronts cycles.

And when cycles are confronted, truth collides with trauma. You will begin to see the lies your survival depended on—and realize they are no longer necessary. The chaos you called normal, the patterns you excused as personality, the pain you thought was permanent—all of it is being recalled and replaced.

GOD is not negotiating with your brokenness. He is curing it. This kind of healing will disrupt generational familiarity. It will require you to grieve what raised you and still choose what will heal you. But peace is worth the war. And truth is the only remedy for a soul conditioned by falsehood.

GOD is not just bringing you out—He's bringing you back into divine order.

Reflection Question:

What patterns in your life feel like they were passed down—and not confronted?

Challenge:

Ask HOLY SPIRIT to show you one area where you've adapted instead of healed. Ask for divine truth to be revealed and for peace to stabilize your response.

____

____

____

____

____

____

PROPHETIC DECREE

I renounce every generational curse, habit, and pattern that tried to anchor me to instability and dysfunction. I break covenant with cycles that trained me to adapt to survival instead of receiving healing. I declare that I will not repeat what was modeled—I am the interruption and the reversal.

I reject emotional instability, spiritual confusion, and inherited ways of thinking that deny the peace, truth, and order of GOD. I call in the success of stability. I call in strength of mind. I call in emotional grounding and spiritual clarity.

The promise in Jeremiah 33:6 is mine. I receive health, healing, and the abundance of peace, security, stability, and truth. I will not build my future on broken foundations. I surrender my survival habits and step fully into restoration.

Within my DNA is a supernatural strain of the Almighty. I carry divine power to transition from what was to what HE said will be. I am not bound by bloodline failures—I am aligned with the blood of JESUS CHRIST, and that blood rewrites everything.

I am not fragile. I am fortified. I am not broken. I am becoming.

I am healed and whole in JESUS CHRIST name, Amen.

This decree was influenced by:
Jeremiah 33:6; John 14:27; Romans 12:2; 2 Corinthians 5:17; 1 Peter 1:18–19

Thoughts:

____

____

____

____

____

Thoughts:

Thoughts:

DAY THREE

Psalm 147:3 KJV

He healeth the broken in heart, and bindeth up their wounds.

Let's Talk About It:

Heartbreak doesn't just come from romance. It comes from life. From people who promised and didn't deliver. From moments you thought would be different. From expectations that shattered on impact. From family, from friends, from systems, from silence.

But GOD doesn't dismiss it. He heals it. Not with shallow platitudes—but with surgical precision. He binds up wounds that no one else sees. And the places where you've been bleeding out emotionally are now on Heaven's triage table.

And when GOD steps in as the Master Surgeon, He doesn't just stop the bleeding—He targets the root. He doesn't wrap trauma in religious language—He restores identity at the source. There are things in you that broke before you had the words to explain the pain. And yet, even those unnamed aches are seen by the One who shaped your heart. The HOLY SPIRIT is not overwhelmed by the intensity of your internal injury. In fact, He's mobilizing healing to every place where time paused, where trust died, and where your voice went silent.

This is the season where divine hands go where human help couldn't. Where Heaven rewrites the diagnosis. Where grief is not rushed, but it is not permitted to reign. GOD is not just comforting you—He is reconstructing you. And what comes out of this healing won't be a softer version of you—it will be a stronger one, sealed by glory and sustained by truth.

Reflection Question:

What wound have you hidden so well that even you forgot it was still bleeding?

____

____

____

____

____

Challenge:

Ask HOLY SPIRIT to help you trace where your heart broke—and invite Him to begin the healing. Let Him pull the memory to the surface not to relive it, but to release it.

PROPHETIC DECREE

LORD, if there was any abuse—known or unknown, seen or unseen—I surrender it into Your hands. I give You permission to reveal what may have been buried and to heal what I didn't know was broken. I call up childhood wounds, mishandled moments, misunderstood trauma. I surrender what shaped me outside of Your love.

I renounce every lie connected to unhealed heartbreak. I speak healing over the abandoned places. I call for divine wholeness to revisit the moments that broke my spirit. I sever every soul tie and memory tethered to pain. I receive supernatural therapy by the HOLY SPIRIT. I invite the Great Physician to stitch what trauma tore.

I rise as one no longer living with silent wounds. I declare healing from the inside out.

I place a demand on divine justice. I call back every fragmented piece of my soul scattered through trauma, shame, and neglect. I issue a spiritual subpoena to suppressed pain, to come into the light—not to torment, but to be transformed. I reject every coping mechanism that disguised my wounds as strength. I repent for the times I mislabeled survival as healing. I no longer normalize what Heaven never called mine.

I invoke the covenant of healing written in the BLOOD. I stand under the government of peace and submit every memory to the LORDSHIP of JESUS CHRIST. Let the anointing go back to the places where my voice was silenced, where innocence was stolen, and where identity was fractured. Let the oil flow into places my words never reached. Let what I couldn't remember be redeemed—not to relive it, but to reclaim what it tried to bury.

I speak to every nervous system response, every emotional scar, and every mental pattern created by pain—and I command a divine reset. I declare that trauma will not instruct my decisions, fear will not anchor my relationships, and memory will no longer interrupt my movement.

Today, I come out of hiding. I walk out of the shadows of silence. I rise—not as the sum of my pain—but as evidence of HIS power.

I am healed and whole in JESUS CHRIST name, Amen.

This decree was influenced by:
Psalm 147:3; Isaiah 61:1; Ezekiel 36:26; Luke 4:18

Thoughts:

Thoughts:

Thoughts:

DAY FOUR

Isaiah 60:1 KJV

Arise, shine; for thy light is come, and the glory of the LORD is risen upon thee.

Let's Talk About It:

There's a version of you that the enemy has tried to bury under shame, failure, and fatigue. But this is not your ending. This is your rising.

"Arise" is not a suggestion—it's a summons. And "shine" isn't about perfection—it's about presence. The light has come. You don't have to manufacture it. You don't have to qualify for it. You just have to respond.

Because when GOD calls you to rise, He's not asking for a performance—He's activating your position. The glory rising upon you is not cosmetic—it's covenantal. This is about alignment, not applause. You are not shining because you have it all together—you are shining because His presence is resting on you. And that light reveals, revives, and reassigns.

There is a divine order to this rising. You are being called up to occupy what darkness tried to deny. You are being summoned into visibility—not to entertain, but to govern. Not to be liked, but to be light. This is your moment to come out from hiding, not because you're flawless, but because you're chosen. Shame cannot stay where glory rests. And fatigue must bow to the fire that now fuels you.

So rise—not with fear, but with fire. Shine—not for attention, but as evidence. The glory of the LORD is not just visiting you—it is rising in you.

Reflection Question:

Where have you dimmed your light to make others comfortable or to avoid exposure?

____

____

____

____

____

Challenge:

Ask HOLY SPIRIT what rising looks like for you in this season. Is it speaking? Is it forgiving? Is it obeying what He already told you to do? Then do it.

____

PROPHETIC DECREE

I renounce the lie that I am too late to rise. I reject the narrative that I missed my moment. I silence the voice of shame that tells me to shrink. I declare that the glory of the LORD is upon me—not because of my performance, but because of His purpose.

I arise into identity. I shine because He has lit me with purpose. I call forth every buried gift, every silent calling, every dormant dream. I declare activation and alignment with what GOD originally spoke.

I refuse to blend in when I was born to break through. I receive the oil, the strength, and the strategy to arise in truth.

I now issue a divine reversal of every delay—natural or spiritual—that attempted to hold me back. I recover lost time by the redemptive acceleration of the LORD. I speak to the timelines of destiny and command realignment. I break every soul tie to seasons that kept me bound to false narratives. I shake off the garments of fear, hesitation, and invisibility. I am not behind—I am right on schedule in the calendar of the Kingdom.

I rise under divine orders. I shine with Heaven's clarity. The anointing on my life is not waiting for man's approval. It is backed by GOD's announcement. I partner with the glory that is already upon me. I synchronize with the sound of the Spirit calling me out of delay, out of doubt, and into dominion. I don't step timidly—I step with governmental weight.

Every system that once silenced me is now subject to my sound. Every space that once ignored me will now witness divine distinction. I call my voice out of captivity. I command the fire on my life to intensify. I legislate a season of visibility, credibility, and fruitfulness—not for fame, but for fulfillment of purpose.

What was once dormant is now divine. What was once buried is now burning. I rise as light in dark places and truth in confused spaces. I rise—not just healed—but heralding the next move of GOD.

I am healed and whole in JESUS CHRIST name, Amen.

This decree was influenced by:
Isaiah 60:1; Matthew 5:14–16; 2 Timothy 1:6; Romans 13:12

Thoughts:

Thoughts:

DAY FIVE

Romans 5:5 KJV

And hope maketh not ashamed; because the love of GOD is shed abroad in our hearts by the HOLY GHOST which is given unto us.

Let's Talk About It:

Disappointment is real and it is raw. It has the ability to make you shrink back and live without expectation. If I don't expect much, I won't get hurt. We don't always say it out loud, but we've lived like it.

Even as believers, disappointment has shaped how we pray, how we expect, and how we hope. I'm not going to tell you that being in CHRIST means you'll never face disappointment—but when we allow GOD's love to enter in, it changes everything. Hope rooted in that kind of love will never leave you empty.

Because hope rooted in His love is not wishful thinking—it's covenant expectation. It's not a gamble—it's grounded. It does not rise and fall based on outcome, but remains steady because it flows from GOD's character, not our circumstances. Disappointment doesn't just affect our hearts—it attacks our theology. It makes us question whether GOD is still good, whether He still hears, whether He still cares.

But Romans 5:5 dismantles that lie. The HOLY GHOST does not just visit us—He pours love into us. And that kind of saturation leaves no room for shame, no space for silent despair. When the love of GOD fills the places disappointment emptied, you rise again—not just to believe, but to expect boldly. Hope becomes holy again. Safe again. Whole again.

This is the healing of hope. Not that you'll never hurt, but that your expectation will no longer be enslaved to fear.

Reflection Question:

Where have you allowed past disappointment to silence your expectation of GOD?

____

____

____

____

____

Challenge:

Ask HOLY SPIRIT to reconnect you to the Word of GOD concerning your hope. Ask Him to highlight one promise you gave up on—and speak life over it again.

PROPHETIC DECREE

I renounce the voice of disappointment that has talked me out of expecting full healing. I silence the inner critic that has shamed me for still believing. I declare that hope is rising in me—not fantasy, not avoidance, but real, anchored, Holy-Spirit-fueled hope.

I disconnect myself from the discouragement that becomes disappointment's twin and ask for help from the Rock that is higher than I.

I reconnect with the Word of GOD which is guaranteed to accomplish what it is sent to complete. I stand convinced that GOD who began the work will complete it. I recall His record. I will not shrink back. I will not be silent. I will hope with my whole heart, because the love of GOD will never fail me.

By apostolic authority, I decree the dismantling of every false conclusion built by disappointment.

I call every unanswered prayer and unresolved pain into divine review under the courtroom of Heaven. I declare that the verdict has not changed: I am still chosen, still called, and still a candidate for healing and wholeness.

I legislate divine hope to invade every barren expectation. I command a resurrection of belief where delay buried my confidence. I summon every word GOD ever spoke concerning me into active circulation. Let the archives of prophecy be reopened. Let the oil of remembrance flow over my hope. Let what was whispered in the dark rise up as strategy in the light.

I reject the theology of disappointment that says GOD does not move anymore. I reject the passive doctrine of survival that tells me to settle for "less than" in the name of wisdom. My hope is not in outcomes—it is in the GOD who commands outcomes. I declare a holy defiance against despair. I cast off hesitation and put on the garment of confident expectation.

This hope will not embarrass me. This hope is not hollow. This hope is not emotional—it is eternal. The HOLY GHOST Himself is pouring love into the cracks disappointment created. Every place where pain lived too long is now under divine renovation.

I prophesy over my heart: beat with boldness again.

I prophesy over my prayers: flow with faith again. I prophesy over my history: bow to the GOD who still writes the end.

I am healed and whole in JESUS CHRIST name, Amen.

This decree was influenced by:
Romans 5:5; Isaiah 55:11; Philippians 1:6; Hebrews 10:23

Thoughts:

Thoughts:

Thoughts:

DAY SIX

Mark 11:23 AMPC

Truly I tell you, whoever says to this mountain, Be lifted up and thrown into the sea! and does not doubt at all in his heart but believes that what he says will take place, it will be done for him.

Let's Talk About It:

Why did JESUS say mountain? He could've chosen anything—a stone, a wall, a stumbling block. But He chose something big, looming, intimidating. That wasn't poetic—it was prophetic. He was talking to people who understood what it felt like to face something massive, immovable, and humanly impossible.

And so do we. Your "mountain" might be your finances. Your health. Your single status. Your family history. Your past. It might be the weight of everything you've survived. But here's what JESUS makes clear: you don't have to climb it. You don't have to carry it. You have to speak to it.

Faith doesn't deny the size of the mountain. Faith just refuses to let the mountain have the final word. And this is where your healing begins—not when the mountain disappears, but when you realize it's an opportunity. An opportunity to use your voice. An opportunity to magnify the LORD. An opportunity to see the supernatural.

You're not speaking from fear—you're speaking to it. And when you speak in faith, something miraculous happens. The mountain moves. Not because you're loud, but because you're aligned.

The same sea that swallows mountains is the same one JESUS invites us to walk on. That's not coincidence—that's covenant. What looked like your biggest problem is now about to become the stage for your greatest miracle.

Reflection Question:

What has been your "mountain"? What have you allowed to feel bigger than GOD?

Challenge:

Ask the HOLY SPIRIT to show you what mountain you've accepted as permanent. Ask Him to reveal how to speak from alignment—so that your words echo Heaven's will.

____

____

____

____

____

____

PROPHETIC DECREE

I lift my voice in faith. I do not deny the mountain—but I deny its authority to stay. I speak to every towering obstacle in my life and command it to move. I declare that what seems impossible is bowing to the name of JESUS CHRIST. I will not circle this mountain again.

I renounce every lie that made me feel powerless. I silence every voice of doubt. I reject the fear that told me I couldn't speak boldly. I refuse to let brokenness or status define me—I am defined by Heaven.

I come into divine alignment. I choose to see this mountain not as a threat—but as an opportunity to magnify the LORD. I make GOD bigger in my mind than any obstacle in my path. I choose to exalt His power over my problem. I refuse to allow anything in my life to loom larger than the GOD who made me.

I speak to the mountain, and I command it to be cast into the sea. The same sea that tried to drown me will now hold what tried to block me. I will walk in supernatural authority. I will not be overwhelmed—I will overcome. I do not drown. I walk on water. I declare that the mountain is not my end. It is my turning point.

Thank You, LORD, for teaching me to use my voice, for giving me power through faith, and for leading me into healing by way of victory. I am healed and whole in JESUS CHRIST name, Amen.

This decree was influenced by:
Mark 11:23; Matthew 14:29; Psalm 34:3; Isaiah 40:4; Zechariah 4:7

Thoughts:

Thoughts:

DAY SEVEN

Psalm 61:2 AMP

From the end of the earth will I cry to You, when my heart is overwhelmed and fainting; lead me to the rock that is higher than I [a rock that is too high to reach without Your help].

Let's Talk About It:

There are moments when even saying "I'm overwhelmed" feels like an understatement. You've been holding it together with tape and prayer. You've been showing up while silently falling apart. You've gotten so used to functioning on fumes, you've almost convinced yourself this is normal. Yet, you need help!

This is what David was crying out for. He knew that in that moment nothing and no one could help him outside of GOD. This verse makes it plain: Only GOD has the capacity to lift you to where you need to be. People can care about you. They can pray for you. But they can't carry you to the Rock. That's GOD's job —and He's not outsourcing it. There are results that only GOD can give and David fully recognized that. Are you in need of results that only GOD can give?

This verse shows us something vital: part of our restoration comes from expecting correctly. Some of the people you're angry with didn't fail you—they just didn't have the capacity to be what you needed. That hurt you felt when your parents couldn't hold you right, or when your spouse couldn't support you, or when the mentor didn't come through? It wasn't rejection—it was limitation. And in some cases, it was divinely allowed.

We must stop expecting people to fill places only GOD can occupy. The Rock is not your spouse. It's not your church or leader. It's not your best friend. The Rock is JESUS—the Chief Cornerstone. The one who was rejected—don't you reject Him too. Because when GOD is your Rock, He cannot be replaced.

Reflection Question:

Is there someone you need to forgive and release because your expectations were beyond their capacity?

____

____

____

____

____

____

Challenge:

Ask HOLY SPIRIT what place in your life—where you've been overwhelmed—He's asking you to come to the Rock with His help. Write it down, release it, and lean into His lifting.

PROPHETIC DECREE

LORD You are my help! You are my rock! Life has overwhelmed me beyond words. I need Your Help. I surrender my emotions to You and ask for clarity and balance in this moment. I have the mind of CHRIST. I have the thoughts, feelings and purpose of His heart. I am not forsaken or forgotten.

I reject every weight of overwhelm that I've been carrying in my own strength. I release every false expectation I've placed on people. I forgive those who couldn't be what I needed. And I release them from the weight of my disappointment. I recognize that they didn't have the capacity—and that's not their fault.

Lift me to the place that I cannot reach without You. Be my strength, my support, and my stronghold. I come out of agreement with striving, survival, and self-reliance. I lean into Your help. I submit to Your rescue. I choose You—my Rock and my Redeemer.

Thank You that I am not carrying this alone. Thank You for raising me up and rebuilding what was too heavy for me. Thank You for being the GOD who sees, who lifts, who leads. I am not overwhelmed. I am held. I am not breaking—I am being built. I am healed and whole in JESUS CHRIST name, Amen.

This decree was influenced by:
Psalm 61:2 AMP; Isaiah 41:10; 1 Corinthians 2:16; Matthew 11:28–30; 1 Peter 5:7

Thoughts:

Thoughts:

Thoughts:

DAY EIGHT

1 Peter 5:10 AMPC

And after you have suffered a little while, the GOD of all grace [Who imparts all blessing and favor], Who has called you to His [own] eternal glory in CHRIST JESUS, will Himself complete and make you what you ought to be, establish and ground you securely, and strengthen [and settle] you.

Let's Talk About It:

This Scripture reveals four things GOD promises to do Himself: complete you, establish you, strengthen you, and settle you.

Let's focus on that last word—settle.

In our culture, settling sounds like weakness. The world trains us to never settle—especially if we want success. But GOD's definition of "settle" isn't a downgrade; it's divine grounding.

Here's what the word settle really means:

- To establish or secure permanently
- To become fixed, resolved, or established
- To take up an ordered or stable life
- Even to conceive

Settling is not where you lose momentum—it's where you gain divine foundation. It's where instability breaks and strength is secured. Settling doesn't mean shrinking. It means being rooted, stable, and safe in GOD's will.

So what does this mean for us? It means that you won't always feel like you're floating. The wilderness won't last forever. The constant motion won't last always. Eventually, there comes a moment when the GOD of all grace puts His foot down in your life and says, This is the place. You will be established here.

It won't be forced. It won't be fake. It will be fixed in Him. Settling is not an end—it is your true beginning. It's what He's been preparing you for all along.

Reflection Question:

Where does your life feel most unsettled or unstable?

Challenge:

Ask HOLY SPIRIT to highlight the specific area He wants to settle in your life. Speak over it with bold agreement and declare His Word until your feet feel firm.

____

____

____

____

____

____

____

PROPHETIC DECREE

I renounce the cultural mindset that taught me settling is weakness. I break agreement with instability, vagabond patterns, and wandering spirits. I declare I will be grounded, anchored, and planted in the will of GOD.

I refuse to romanticize restlessness. I reject the idea that movement always means progress. I declare that divine stability is my portion, and the grace to be settled is already at work in my life. I will no longer grieve endings that were actually entrances.

Thank You, LORD, for completing what You started in me. Thank You for establishing my footing, strengthening my frame, and settling me into my Kingdom purpose. The winds of uncertainty do not define me—Your Word does. You are the GOD who plants and prospers. I am not lost. I am located. I am not shaken. I am secure.

I am healed and whole in JESUS CHRIST name, Amen.

This decree was influenced by:
1 Peter 5:10; Psalm 40:2; Philippians 1:6; Hebrews 12:28; Isaiah 33:6

Thoughts:

Thoughts:

Thoughts:

DAY NINE

Psalm 23:3 KJV

He restoreth my soul: He leadeth me in the paths of righteousness for His name's sake.

Let's Talk About It:

What is the enemy after? Your soul. Your soul is the battlefield. Not your body. Not your money. Your soul—the seat of your mind, will, and emotions—is what the enemy targets, twists, and wears down. He can't beat you, so his strategy is to wear you down until you give up.

This is exactly why GOD promises to restore it. The weariness. The fatigue. The utter depletion. "He restoreth my soul..." This isn't poetic comfort—it's prophetic confrontation. It means there's something in you that has become fractured, disoriented, exhausted, or fragmented. And GOD is not avoiding it—He's rebuilding it.

The fight for your soul is the fight for your direction. Because right after restoration, the verse says, "He leadeth me in the paths of righteousness..." The enemy is invested in you taking the wrong steps and the wrong path. And when we are wounded, it's easy to be led astray with a mirage of help. But HOLY SPIRIT will lead and guide us into all Truth.

That means divine restoration doesn't leave you wandering. It sets you back in the right place, on the right path, for the right reason.

And none of it—not the healing, not the guidance, not the covering—is earned. It's given for His name's sake. That means even when you don't feel worthy, even when you don't feel ready, He's still committed to your recovery because His name is on the line. His thoughts toward you are for good and to give you an expected end. His end—which is connected to the good path as identified in Ephesians 2:10.

So if you need to get a snack and take a nap like Elijah, it is okay. Get restored. But know that when you get up, there is work to do—because someone's survival is waiting on your obedience.

Reflection Question:

What is the enemy using to target your soul? What currently has you in unrest?

____

____

____

____

____

Challenge:

Ask HOLY SPIRIT to show you the door that's allowing the enemy access to your mind, will, or emotions. Where have you rested in something outside of GOD? Break agreement with every alliance, comfort, or relationship that drains your soul.

PROPHETIC DECREE

I enter the rest of Your love by faith. I choose to allow Your rest to capture my soul and command every resistance to Your will be destroyed. I renounce every false alliance and counterfeit comfort I've leaned on that has drained my soul. I close every gate the enemy has used to access my mind, will, and emotions. I sever ties with unreciprocated relationships, manipulative voices, and spiritual leeches. I declare that my soul belongs to GOD, and I refuse to rent space to anything unclean.

I evict confusion. I break soul fatigue. I come out of alignment with every cycle that warred against my restoration. I declare that restoration is not optional—it is active. The breath of the LORD is breathing strength into my being. I will run and not be weary and walk and not faint.

I receive divine boundaries, divine wisdom, and divine positioning. My soul will no longer be swayed by emotional warfare or false priorities. I am not scattered—I am stabilized. I am not depleted—I am replenished. I am not confused—I am clear. I know the path I should take and shall come forth as pure gold.

I am healed and whole in JESUS CHRIST name, Amen.

This decree was influenced by:
Psalm 23:3, Isaiah 40:31, John 16:13, Ephesians 2:10, Psalm 103:1–5

Thoughts:

Thoughts:

Thoughts:

DAY TEN

2 Samuel 4:4 KJV

And Jonathan, Saul's son, had a son that was lame of his feet. He was five years old when the tidings came of Saul and Jonathan out of Jezreel, and his nurse took him up, and fled: and it came to pass, as she made haste to flee, that he fell, and became lame. And his name was Mephibosheth.

Let's Talk About It:

What happens when life drops you?

Mephibosheth didn't ask for what happened to him. He was dropped. And he was never the same again.

There are moments that break us—not because of something we did, but because of something someone else did. Our lives have had to hang in tandem because of something we couldn't control.

Abuse you didn't ask for.
Betrayal you didn't deserve.
A boss who didn't like you and made your life miserable for no reason.
A moment where no matter how hard you tried, nothing could fix what was falling apart.
You were dropped. And now you've had to live with the consequences of something that wasn't your fault.

But here's what makes this story prophetic: Even though Mephibosheth was dropped, he was not forgotten.

David remembered the covenant he made with Jonathan—Mephibosheth's father. He sent for him, restored his position, and gave him a permanent seat at the king's table.

That is the picture of GOD's heart for you. You may have been dropped. You may feel overlooked, unheard, or wounded in a way that still limits you. But you have not been forgotten. There is a covenant over your life—a promise tied to your name.

The KING remembers. And when GOD remembers His word concerning you, He restores not based on your ability—but based on His integrity. Mephibosheth was still given a seat. He was still restored. And you will be too.

But here is where the work comes in: Your heart still needs to remain pure. You can't afford to become bitter. You have to forgive those who may have dropped you—those that left, abandoned, or rejected you. Their treatment of you cannot determine your outlook. How someone treats you is not up to you—but how you respond is.

The enemy wants you to become bitter so you forfeit the promise. GOD tells us not to come to Him if we have aught against a brother. We're commanded to get it right first. If we don't forgive, we can't be forgiven—and none of us can afford that.

Though life has dropped you, unlike Mephibosheth you have the ability to get up and walk again. Like the crippled man in Scripture, pick up your bed and walk. Walk in love and forgiveness—ultimately walking in healing and wholeness. Now that is powerful!

Reflection Question:

Where in your life are you still feeling the effects of being dropped?

____

____

____

____

____

Challenge:

Ask HOLY SPIRIT to reveal where bitterness may be trying to hide and block your restoration. Renounce it and release those who could not carry you. Call yourself into wholeness and pick up your mat.

PROPHETIC DECREE

I am not forsaken or forgotten. I am remembered by the King. I am loved by the King—and I respond by loving Him back.

LORD, if there was any abuse—known or unknown, seen or unseen—I surrender it into Your hands. I give You permission to reveal what may have been buried and to heal what I didn't know was broken. I choose to forgive. I renounce the pain of being dropped. I call back every fragmented piece of my identity that was wounded through mishandling, abandonment, or betrayal.

Thank You for being the GOD who restores and remembers. Thank You for the covenant that still stands over my life. I receive restoration and a seat at Your table. I will not let offense abort my inheritance. I walk in love. I walk in freedom. I walk in healing and wholeness.

I am healed and whole in JESUS CHRIST name, Amen.

This decree was influenced by:
2 Samuel 4:4, 2 Samuel 9:1–7, Luke 5:24, Matthew 6:14–15, Isaiah 61:7

Thoughts:

Thoughts:

Thoughts:

DAY ELEVEN

Joel 2:25–26 KJV

And I will restore to you the years that the locust hath eaten, the cankerworm, and the caterpiller, and the palmerworm, My great army which I sent among you. And ye shall eat in plenty, and be satisfied, and praise the name of the LORD your GOD, that hath dealt wondrously with you: and My people shall never be ashamed.

Let's Talk About It:

You didn't just lose time—you lost hope. It wasn't just one thing. Joel mentions four devourers—locust, cankerworm, caterpillar, and palmerworm. That's not poetic repetition—it's prophetic precision. It means the attack came in waves. It wasn't just one disappointment, one delay, one crisis. It was layered. And at times, you felt bombarded. Just as you were healing in one area, another broke down. Just as you stood up, something else knocked the wind out of you.

But here's the promise: restoration is coming. And not just in pieces, but in purpose. GOD has a plan for you—and He is invested. Are you? You will eat in plenty and be satisfied. You won't just survive—you'll thrive. The promise is abundant life.

And the second part of the promise is just as powerful: you will never be ashamed. What tried to publicly humiliate you will become the platform for GOD's public restoration.

This is not just healing—it's Heaven's announcement that your drought is over, and your story is not ending in shame.

Reflection Question:

Where in your life do you feel like time has been stolen, delayed, or devoured?

Challenge:

Ask HOLY SPIRIT to show you the specific area He wants to restore first. Partner with Him in obedience and agreement. Speak over that place what the enemy tried to erase.

PROPHETIC DECREE

I bring to You the loss of hope and expectation. I surrender the broken places. I renounce every lie that says it's too late. I break agreement with every voice that said "it's over." I call back the years that were devoured—every locust, cankerworm, caterpillar, and palmerworm that came to steal from me has been issued a cease-and-desist in the Spirit. Restitution is now!

I decree divine restoration over my timeline. I partner with my Kairos moment. Every wasted year, every stolen opportunity, every hidden or hijacked assignment is now being returned with interest. The years that were swallowed up in survival are being transformed and used for my optimal utilization.

I thank You, LORD, that I will be satisfied. I thank You for the public restoration You are orchestrating. My shame will not have the final word. Your wonder-working power is rewriting my story in front of those who thought it was finished.

The heavens are aligning with this decree. I will see the evidence in my body, my soul, my money, my ministry, and my legacy. Restoration is my portion—and I receive it.

I am healed and whole in JESUS CHRIST name, Amen.

This decree was influenced by:
Joel 2:25–26, John 10:10, Isaiah 61:7, Psalm 126:5–6, Romans 8:18

Thoughts:

Thoughts:

Thoughts:

DAY TWELVE

Psalm 141:3 AMPC

Set a guard, O LORD, before my mouth; keep watch at the door of my lips.

Let's Talk About It:

Some of us grew up in environments where we were silenced before we even learned how to speak clearly. We were told not to talk back, not to express feelings, not to question authority. And so, we learned early: Your voice doesn't matter.

For others, we did speak—but only from wounds. So we lashed out to be heard, we raised our voices to feel seen, we vented because we didn't feel validated. But what happens when your voice is shaped by trauma? By pain? By a desperate need to be understood?

This verse—Psalm 141:3—isn't about silencing your voice. It's about sanctifying it. It's about healing the places that taught you to weaponize your words and learning to speak with divine restraint and wisdom. Because not every silence is weakness, and not every loud voice is strength. Sometimes the greatest power is learning how and when to speak.

GOD is not asking you to silence your voice. He's asking you to surrender it. To stop using your words as swords when they were meant to build altars. To allow the HOLY SPIRIT to filter your emotions before you form sentences. To let Heaven sanctify your sound.

Reflection Question:

Have you been speaking from a healed place—or a hurting one?

Challenge:

Ask HOLY SPIRIT to show you the emotion that most often hijacks your voice. Practice holding space with GOD before speaking, allowing Him to filter your heart before you respond.

PROPHETIC DECREE

LORD, I surrender my voice to You. I place my mouth on the altar and ask You to sanctify it. Set a guard, O LORD, over my mouth; keep watch over the door of my lips. I no longer speak from pain, offense, or the need to prove myself. I renounce the identity shaped by being silenced, dismissed, or misunderstood.

I sever ties with every emotional trigger that taught me to weaponize my words. I reject every false narrative that made me think I had to be loud to be heard or silent to be accepted. I will no longer speak to survive—I speak to align. I speak with clarity. I speak with purpose. I speak with restraint when needed and with boldness when commanded.

LORD, heal the parts of me that are still grieving the moments where I was silenced. Restore the purity of my voice. Let my words reflect Heaven's rhythm and not hell's reaction. I prophesy that my mouth will build, bless, and birth what glorifies You.

I thank You that my voice is not lost—it is being recalibrated. I thank You that my silence will no longer be passive, but powerful. I speak as one who has been healed. I speak as one who has been made whole.

I am healed and whole in JESUS CHRIST name, Amen.

This decree was influenced by:

Psalm 141:3, Proverbs 18:21, Ecclesiastes 3:7, Isaiah 6:6–7, Luke 12:12, James 1:19

Thoughts:

Thoughts:

Thoughts:

DAY THIRTEEN

Colossians 2:10 KJV

And ye are complete in Him, which is the head of all principality and power.

Let's Talk About It:

Where did you first learn you weren't enough? Maybe it wasn't said outright, but something communicated it—through silence, criticism, rejection, comparison. And since then, life has offered endless substitutes for true completion.

We've sought completion in love, success, busyness, appearance, approval—and still felt empty. Why? Because none of those things were designed to complete us. They were designed to reflect GOD, not replace Him.

Paul's words in Colossians are a confrontation to every insecurity: "Ye are complete in Him." It's not poetic—it's positional. Completion is not a goal—it's your starting point in CHRIST.

Completion in GOD means you stop striving for worth and start receiving it. It means you stop attaching your value to what's missing and anchor it to what cannot be taken.

Yes, you still may grow, mature, or heal—but none of that means you're lacking. You are being shaped, but you are not broken. The enemy wants you to keep looking for completion in what has no power. But you are already filled by the One who reigns over all.

The question isn't *if* you are enough. The question is—*do you believe Him when He says you are?*

Reflection Question:

What substitute have you been depending on to feel complete—and is it still leaving you empty?

____

____

____

____

____

Challenge:

Ask the HOLY SPIRIT to show you where your identity has been rooted in insufficiency. Break agreement with those mindsets and declare out loud: "I am complete in CHRIST."

____

____

PROPHETIC DECREE

I declare that I am complete in CHRIST, who is the head of all principality and power. I repent for seeking identity, security, and validation in anything or anyone other than You, GOD. I renounce the lie that said I had *to do more, be more, or have more* to be worthy. I renounce the false completion I tried to find in relationships, sex, food, work, social media, performance, and perfectionism.

I reject the empty promises of temporary fillers. I sever emotional soul ties and mental agreements with everything that made me feel less than, broken, or insufficient. I declare that I lack nothing in You. I am not a deficit—I am a dwelling place of divine fullness. I renounce insecurity, self-hatred, and false comparisons. I break covenant with the spirit of rejection, fear of abandonment, and shame.

I am not striving—I am seated. I am not fractured—I am filled. I am not overlooked—I am known. CHRIST is my completion, and no other source can take His place in my soul.

Thank You for healing the cracked mirrors that distorted my view of myself. Thank You for anchoring my value in Your image and not in the approval of man. Thank You for being the GOD who completes and restores.

I am whole. I am rooted. I am enough in You.
I am healed and whole in JESUS CHRIST name, Amen.

This decree was influenced by:
Colossians 2:10, Ephesians 1:23, Psalm 23:1, John 1:16, Isaiah 43:4

Thoughts:

Thoughts:

Thoughts:

DAY FOURTEEN

Proverbs 13:4 KJV

The soul of the sluggard desireth, and hath nothing: but the soul of the diligent shall be made fat.

Let's Talk About It:

Slothfulness isn't always laziness—it can be the residue of trauma. When life hits hard, it can drain not just your energy, but your drive. You used to be motivated. You used to dream. But disappointment has a way of convincing you that discipline isn't worth it. You stop starting. You stop finishing. You stop believing that what you do will matter.

But that lie must be broken.

This Scripture reminds us that desire without diligence produces nothing. Craving without commitment leads to frustration. Your soul might be longing for breakthrough, but if your hands refuse to move, nothing will change. That's not just natural laziness—that's a spiritual war. And the only way to win is to fight back with order, structure, and follow-through.

You weren't created to quit. You weren't saved to sit. It's time to confront the emotional clutter that's been disguising itself as tiredness. The dream isn't dead—it's waiting on your diligence.

Reflection Question:

What has disappointment or delay caused you to stop doing? What task, assignment, or promise have you allowed to sit because of emotional fatigue?

Challenge:

Ask HOLY SPIRIT to show you the root of your slothfulness or inconsistency. Partner with Him by re-engaging one thing you dropped—and commit to taking one step forward today, not based on how you feel, but what you believe.

PROPHETIC DECREE

Father, in the name of JESUS CHRIST,

I come before You with full awareness that slothfulness is not just a habit—it is a spirit that wars against purpose, and I refuse to partner with it any longer. Your Word says, "The soul of the sluggard craves and gets nothing, while the soul of the diligent is richly supplied" (Proverbs 13:4). I reject every craving without action. I renounce every desire without discipline. I break agreement with every cycle that delays obedience.

LORD, I repent for every moment I've allowed feelings to override faith. For every day I've mismanaged. For every task I postponed under the guise of "not ready." For every gift You've given that I buried in fear, fatigue, or disorganization. Your Word declares that the diligent shall rule (Proverbs 12:24), and I declare: I will no longer be ruled by excuses.

Stir up in me a supernatural appetite for order, structure, and follow-through. Awaken the mind of CHRIST within me—the mind that finishes what was started. I command my body to come into submission. I call my emotions into alignment. I call my time into purpose. I dismantle the laziness that hides behind busyness and expose the slothfulness that dresses up in distraction.

BREATHE on my routine. BREATHE on my calendar. BREATHE on my waking and my working. I declare: I am being re-trained in the Spirit. I am rising early to meet with You. I am stewarding what You've placed in my hands. I long for discipline—not just for results, but because I want to be ready for what You are releasing.

I prophesy over myself: I am focused. I am consistent. I am not driven by feeling—I am driven by faith. And I welcome the divine discomfort that comes with growth. I am healed and whole in JESUS CHRIST name, Amen.

This decree was influenced by:
Proverbs 13:4, Proverbs 12:24, Hebrews 12:11, Colossians 1:29

Thoughts:

Thoughts:

Thoughts:

DAY FIFTEEN
2 Corinthians 10:5 AMPC

[Inasmuch as we] refute arguments and theories and reasonings and every proud and lofty thing that sets itself up against the [true] knowledge of GOD; and we lead every thought and purpose away captive into the obedience of CHRIST (the MESSIAH, the Anointed One).

Let's Talk About It:

Not every thought you have is true. Some of them were planted by trauma, others by disappointment—and some were handed to you by people who didn't know your worth. But here's the truth: negative self-talk and thoughts of fear, failure, and unworthiness are not from GOD. They are against the knowledge of GOD and must be cast down.

GOD didn't tell you that you were unqualified, broken beyond repair, or too late. The thoughts that repeat lies like "I'm not good enough" or "It will never happen for me" must be arrested. Your mind is not a playground for torment—it is a territory that belongs to the LORD. And you are not crazy for needing help mastering your thoughts. You are being trained to war in the Spirit. You are taking your mind back from the enemy.

This is not just about thinking positively. This is about divine alignment. If it didn't come from GOD—it must come down.

Reflection Question:

What thought patterns in your life contradict the truth of GOD's Word about who you are?

Challenge:

Ask the HOLY SPIRIT to expose one recurring lie or toxic thought pattern you've come into agreement with. Write down the truth from Scripture and begin speaking it over your mind.

PROPHETIC DECREE

I declare I have the mind of CHRIST. I have the thoughts, feelings, and purposes of His heart. My mind is His domain and under His authority. I take authority over every voice that contradicts what GOD has said about me. I renounce every lie that I've adopted as truth. I silence the inner critic that has spoken against my identity, my worth, and my future. I command every high thing that has exalted itself against the knowledge of GOD to come down.

I will no longer agree with the voice of fear, shame, or unworthiness. I pull down negative self-talk, toxic comparisons, and the residue of rejection. I come out of agreement with imposter syndrome, inadequacy, and every mental barrier that has tried to limit me. I cast down imaginations and bring every thought into alignment with CHRIST.

Because I have the mind of CHRIST. I speak truth over my mind. I speak peace over my thoughts. I declare mental clarity, emotional balance, and spiritual discernment. My mind is being renewed, my patterns are being reset, and my soul is being restored. I am healed and whole in JESUS CHRIST name, Amen.

This decree was influenced by:

2 Corinthians 10:5, Romans 12:2, Philippians 2:5, 1 Corinthians 2:16, Psalm 23:3

Thoughts:

Thoughts:

DAY SIXTEEN

Psalm 55:12–14 AMPC

For it is not an enemy who reproaches and taunts me—then I might bear it; nor is it one who has hated me who insolently vaunts himself against me—then I might hide from him. But it was you, a man my equal, my companion and my familiar friend. We had sweet fellowship together and used to walk to the house of GOD in company.

Let's Talk About It:

Some wounds go deeper than others. Not because of what was said or done—but because of *who* did it. The sting of betrayal isn't about the act alone—it's about the proximity. This was someone you trusted. Someone you prayed with. Someone you considered safe. And now, you're carrying the weight of a wound that doesn't make sense.

David doesn't say he was attacked by a stranger—he was wounded by someone he worshiped with. That makes the pain more personal. The trauma of betrayal is often buried under our "I'm fine" masks, but it leaves a spiritual bruise that can affect how we trust, love, and connect.

Forgiveness doesn't mean the pain wasn't real. It means you're releasing the right to carry it. You're refusing to let bitterness rewrite your story. You're choosing freedom over fury. Healing doesn't erase the memory—it disarms its power. And just as GOD forgave you, He's inviting you to forgive them—not for their sake, but for yours.

Reflection Question:

Who hurt you that you haven't fully released? What would it look like to forgive them—even if they never apologize?

____

____

____

____

____

Challenge:

Ask the HOLY SPIRIT to reveal any hidden anger or unresolved betrayal that's hindering your healing. Invite Him into the wound. Speak their name out loud and say, "I release you. I choose freedom."

____

____

____

PROPHETIC DECREE

I come to You recognizing that I am still carrying this wound. I release the need to look strong and ask You to break every embittered root in my foundation so I can truly be strong. I break agreement with every lie that tells me holding on will protect me. I release the pain, the people, and the punishment. I declare that the betrayal I faced will not bleed into my future. I refuse to carry bitterness in my body, in my heart, or in my relationships. Help me heal!

I am not bound by what they did—I am healed by what CHRIST did. I silence the torment of replaying moments I can't change. Though it may be difficult, I choose to forgive and walk in liberty. I declare: my soul will not stay stuck in the past. I release the need for an apology or understanding as to why they did what they did, and I embrace the power of GOD's restoration.

I thank You, LORD, for healing the wound I didn't want to admit. I praise You for severing the soul tie of offense. I declare that my heart is clean, my spirit is whole, and my future is unblocked. I am free from the trap of betrayal. I will not take the bait. I am healed and whole in JESUS CHRIST name, Amen.

This decree was influenced by:

Psalm 55:12–14, Matthew 6:14–15, Hebrews 12:15, Ephesians 4:31–32

Thoughts:

Thoughts:

Thoughts:

DAY SEVENTEEN

Jeremiah 29:11 AMPC

For I know the thoughts and plans that I have for you, says the LORD, thoughts and plans for welfare and peace and not for evil, to give you hope in your final outcome.

Let's Talk About It:

Somewhere along the way, you may have believed a lie: that GOD wasn't really for you. That He loved others more. That your pain disqualified you from His promises. That because of what you experienced, maybe you were being punished. Maybe you weren't chosen. Maybe you were forgotten.

But this Scripture silences all of that. GOD is invested in you. He is not the author of your trauma. He is the architect of your triumph. His thoughts toward you are good—not neutral, not indifferent, not harsh, but good. And He doesn't just have feelings—He has plans.

He's more committed to your success than you are. He's not looking for an excuse to disown you—He's been designing your restoration before you even realized you needed it. So now is the time to break agreement with every lie that said GOD wasn't for you. Because the truth is, you've never been without His intention. And He hasn't changed His mind.

Reflection Question:

What belief or lie have you carried that makes it hard to believe GOD wants you to win?

Challenge:

Ask HOLY SPIRIT to reveal every hidden agreement you've made with the idea that you're unwanted, unloved, or unworthy. Write it out—and then renounce it with the Word.

PROPHETIC DECREE

I am part of a chosen generation. I am accepted in the Beloved. I am not GOD's stepchild—I am a joint heir with CHRIST. I have an inheritance.

I renounce every lie that said I was unwanted, unworthy, or unloved. I silence the voice of shame that tried to twist my past into proof that GOD wasn't for me. I break covenant with every mindset that made me believe I had to earn what GOD already gave.

I announce to every corner of my soul: I am not forgotten. I am not forsaken. I am not being punished—I am being perfected. GOD is invested in my outcome. His thoughts toward me are good, and His plans for me are sure. I have promises in the Word over my life, and the Scripture cannot be broken.

I receive the truth that I am loved, chosen, and seen. I will not fight for acceptance—I live from it. I declare that every setback is being reworked for my good. I lean into Your plans, GOD, not mine. I trust that You are bringing me to the end that You expected—and it is better than I imagined. I embrace Your goodness for me.

I am healed and whole in JESUS CHRIST name, Amen.

This decree was influenced by:
Jeremiah 29:11, Romans 8:17, 1 Peter 2:9, Ephesians 1:6, John 10:35, Romans 8:28

Thoughts:

Thoughts:

Thoughts:

DAY EIGHTEEN

Ephesians 2:10 AMPC

For we are God's [own] handiwork (His workmanship), recreated in CHRIST JESUS, [born anew] that we may do those good works which GOD predestined (planned beforehand) for us [taking paths which He prepared ahead of time], that we should walk in them [living the good life which He prearranged and made ready for us to live].

Let's Talk About It:

The enemy isn't just after your peace—he's after your function. He's not just trying to distract you—he's trying to disconnect you from the good works you were created for. He knows that once you start walking in the paths GOD has already prepared, you'll live the good life that GOD intended—not the life culture demands, but the one Heaven authored. When was "good" first mentioned? In Genesis 1—when GOD created. Every time He completed a portion of His work, He looked at it and called it good. So when Ephesians 2:10 says we were created for good works, it's not casual. It's creative. That means your identity and assignment are directly tied to the purpose in which GOD created you. The "good life" isn't something you build—it's something you align with.

But here's the problem: the enemy saw your potential before you did. So he tried to link you to dysfunction. Through trauma, shame, generational cycles, abuse, delays—he tried to distort the way you saw yourself so you'd never believe you were made for more. He wanted to confuse the path. But the HOLY SPIRIT is your navigation system, and He never loses signal. The paths were prepared beforehand. You don't have to create them. But you do have to choose them.

Reflection Question:

What area of your life feels out of sync with GOD's divine function? What path do you feel drawn to that you've been avoiding?

____

____

____

____

____

Challenge:

Ask the HOLY SPIRIT to reveal the original path He prepared for you in this season. Surrender your plans and align with Heaven's blueprint. Say yes to what GOD already ordained.

PROPHETIC DECREE

I was created for the good life. I am not a coincidence or random. I will not be dysfunctional. I am the handiwork of GOD, created on purpose and for purpose. I silence every lie that told me I had to make life up as I go. I reject the chaos and confusion that attempted to detour my destiny. I break agreement with every trap, trauma, and toxic cycle sent to sabotage my assignment. I recognize and choose GOD's paths.

I receive the truth: I was created for good works. I was built for divine function. The good life is not a myth—it is a promise. I disconnect from every counterfeit path and every pressure that tries to rush me into what I was never designed to carry. I am not here to impress—I am here to fulfill what Heaven wrote. I was created to destroy the works of the enemy. LORD, show me how to destroy them in me first.

HOLY SPIRIT, I lean into Your direction. I trust the path You've prepared. I renounce every voice that said I missed my moment or messed up too badly. I declare: the path is still open, the promise is still intact, and the GOD who planned it is still faithful to perform it. I align with the divine blueprint. I walk with holy confidence. I am healed and whole in JESUS CHRIST name, Amen.

This decree was influenced by:
Ephesians 2:10, Genesis 1:4, 10, 12, 18, 21, 25, 31, Romans 12:2, Jeremiah 1:5, John 16:13

Thoughts:

Thoughts:

Thoughts:

DAY NINETEEN

Psalm 138:8 AMPC

The LORD will perfect that which concerns me; Your mercy and loving-kindness, O LORD, endure forever —forsake not the works of Your own hands

Let's Talk About It:

We've all heard it: "You made your bed, now lie in it." Culture trains us to take full responsibility for our mistakes—and while that may sound noble, it often turns into quiet torment. Regret convinces us that since we messed it up, it's our job to fix it. But what happens when you can't?

About nine years ago, I was driving from Virginia back to Georgia. GOD told me which way to go—and I disobeyed. I got a ticket. I immediately repented for not following His instructions, but the torment didn't stop. For the next thirty minutes of that seven-hour drive, I beat myself up. How am I going to pay this ticket? I kept replaying the moment, trying to figure it out on my own.

That's when the HOLY SPIRIT interrupted my spiraling and said:
"Shoulda, woulda, coulda will all burn in hell."
That meant the mistake was made and couldn't be

changed. But I felt like He was saying—that's not the focus, just the fact.

Then GOD said something that wrecked me:
"I am not just GOD over your perfect decisions—I am GOD in your mistakes."

It wasn't my first inclination to invite GOD into my mess. The thought that He still wanted to help me even if I disobeyed was foreign. That even when I disobeyed, I wasn't disqualified from His desire to still be a present Help in the time of trouble.

Here is the part that humbled me: shortly afterward, someone called and said, *"GOD laid you on my heart—I want to bless you."* And what they gave covered the amount of the ticket.

That's who GOD is. He's not only present in your perfection—He's present in your process. Psalm 138:8 says *He perfects the things that concern us.* Even the things we caused. Even the places we're still trying to clean up.

It's time to break agreement with torment. You are not your mistake. GOD isn't punishing you—He's trying to partner with you. The world says lie in the bed. GOD says pick it up and walk. Your miracle awaits!

Reflection Question:

Where have you been trying to fix what GOD has been waiting to perfect?

Challenge:

Ask the HOLY SPIRIT to reveal one place where regret has kept you stuck in self-reliance. Invite GOD into the mess and ask Him to perfect it for His glory.

PROPHETIC DECREE

LORD, I repent for the moments I tried to fix what only You can perfect. I come out of agreement with regret, shame, and self-punishment. I renounce the lie that I must suffer for my mistakes. I break the torment that tells me I missed it, ruined it, or must now live in lack because of it.

I receive the truth: You are not just GOD over my perfect decisions—You are GOD in my mistakes. You see my missteps and You still stay. You step into the places I tried to handle alone and You bring mercy with You.

I thank You for divine intervention. I thank You that You care about what concerns me. I release the bed of torment and I rise in trust. I pick up my mat and walk into wholeness. I silence every voice of condemnation and embrace the voice of restoration. I am not lost—I am being led. I am not broken—I am being rebuilt.

The heavens are aligning with this decree. I will see divine solutions, unexpected provision, and the mercy of GOD in places I thought were ruined. I am healed and whole in JESUS CHRIST name, Amen.

This decree was influenced by:
Psalm 138:8, Romans 8:28, Psalm 34:18, Hebrews 4:16, and John 5:8

Thoughts:

Thoughts:

DAY TWENTY

Genesis 25:30-31 KJV

And Esau said to Jacob, Feed me, I pray thee, with that same red pottage; for I am faint: therefore was his name called Edom. And Jacob said, Sell me this day thy birthright.

Let's Talk About It:

Desperation can deceive you. That's exactly what happened to Esau. He was hungry—so hungry that he gave up his identity, his inheritance, and his legacy for a moment of relief. And while we may shake our heads at Esau, the truth is we've done it too.

Have you ever been so hungry for something that it created an urgency—an insatiable desire? We've let the urgency of now convince us to forfeit what was meant to be forever. We've traded intimacy with GOD for attention from people. We've given up purpose for popularity. We've compromised integrity for the illusion of security. Why? Because desperation makes you shortsighted.

How many of us have settled for relationships we knew were wrong—just so we wouldn't feel alone? How many of us have used food, sex, drugs, or attention like anesthesia—just to silence a craving that felt unbearable?

How many times have we said, "I need this," when what we really needed was GOD?

Desperation distorts. It rushes you. It whispers that relief is better than righteousness. It makes you forget who you are just to soothe what you feel. But when you numb the pain with compromise, the craving never really leaves—it just deepens. And eventually, you look up and realize you've given up too much.

But the good news is: identity can be restored. Desperation doesn't have to write your story. JESUS CHRIST came to rewrite it—to remind you of who you are and what's already yours. You are not what you settled for. You are not what you gave away in your worst moment. Your birthright is still available—but you must choose it.

Reflection Question:

What decision in your life do you know was made from a place of desperation? Have you let that moment define you?

___

___

___

Challenge:

Ask the HOLY SPIRIT to reveal any moment where you've sold your birthright—your identity, value, or inheritance—for something temporary. Repent and receive restoration.

PROPHETIC DECREE

I am created in the image of GOD. I have purpose and power. I renounce every decision made from desperation. I break covenant with cravings that cost me clarity. I reject every appetite that wars against my identity in CHRIST. I am not what I lost. I am not what I did to survive. I am not what the enemy told me I had to be. I am chosen, called, and set apart.

I will not live in regret. I will not be ruled by shame. I take back every place where desperation tried to rob me of my destiny. I close the door on panic, and I open the door to divine provision. I declare that my hunger will not hijack my inheritance.

HOLY SPIRIT, retrain my appetite. Redirect my desires. Restore my discernment. I receive my identity back. I recover my vision. I reclaim my portion. I am healed and whole in JESUS CHRIST name, Amen.

This decree was influenced by:
Genesis 25:30–31, Romans 8:17, Galatians 3:29, Hebrews 12:16–17

Thoughts:

Thoughts:

Thoughts:

DAY TWENTY-ONE

1 John 3:9 AMPC

No one born (begotten) of GOD [deliberately, knowingly, and habitually] practices sin, for GOD's nature abides in him [His principle of life, the divine sperm, remains permanently within him]; and he cannot practice sinning because he is born (begotten) of GOD.

Let's Talk About It:

This verse is not just about behavior—it's about identity. When you are born of GOD, His nature is in you. That means sin isn't just rebellion—it's identity confusion. It's a temporary denial of who you really are.

At the root of every repetitive sin cycle is a question of worth, belonging, or satisfaction. We run to sin not just out of desire, but out of disbelief—disbelief that GOD can truly satisfy, disbelief that we're really made new, disbelief that we're worthy of holiness.

This is deeper than "doing bad things." This is about agreeing with a lie. Like Peter, who denied CHRIST not because he didn't believe in Him—but because he forgot who he was. That's what sin does: it convinces you that you are less than what GOD already declared. But today, we tear that lie down.

Reflection Question:

Where in your life have you excused sin or unhealthy patterns because you've forgotten who you really are?

Challenge:

Ask the HOLY SPIRIT to reveal any behaviors, cravings, or compromises you've made from a place of broken identity. Let Him speak truth to that place and affirm who you really are. Then renounce the lie and align your life with His DNA.

PROPHETIC DECREE

I am born of GOD. I carry His nature. I reject every lie that tells me I'm not enough or that I need to sin to be satisfied. I break agreement with every identity crisis that has led me into behavior that contradicts who I am in CHRIST. I declare that my body, my mind, and my choices are submitted to the nature of GOD in me.

I renounce the cycles of compromise. I dismantle the false identities I've worn to be accepted, loved, or validated. I repent for every time I turned to sin for comfort. I cast off the shame and confusion that distorted how I saw myself.

I receive the truth: I am a child of GOD. I am not ruled by fear. I am not addicted to false fulfillment. I am not trapped in old habits. The divine nature of CHRIST is alive in me—and I live from that truth. I declare: I am whole. I am free. I am clean.

I am healed and whole in JESUS CHRIST name, Amen.

This decree was influenced by 1 John 3:9 AMPC

Thoughts:

Thoughts:

DAY TWENTY-TWO

Genesis 3:6 AMPC

And when the woman saw that the tree was good (suitable, pleasant) for food, and that it was delightful to look at, and a tree to be desired in order to make one wise, she took of its fruit and ate; and she gave some also to her husband, and he ate.

Let's Talk About It:

The first fall happened because of an appetite. And every fall since has followed the same pattern.

All throughout Scripture, people have either succeeded or failed based on what they fed. Appetite isn't just about food—it's about desires. Eve didn't eat because she was hungry. She ate because she was deceived into thinking something outside of GOD would give her what only He could provide.

And it's still happening today. Some of the appetites we wrestle with didn't come from us. They were awakened through trauma, abuse, molestation, abandonment, or violation. But even if you didn't choose it, now you have to confront it. Why? Because what we feed will grow—and what we starve will die.

I remember one day in prayer for the Body of CHRIST, I had a vision. I saw two creatures fighting—one was huge and muscular, and the other looked weak and emaciated. I asked the LORD what it meant, and He showed me: "That's the war between flesh and spirit. Whatever you feed will live, and whatever you starve will die."

This is not about shame. This is about choice. You are not hopeless or helpless. You are not too far gone. The LORD GOD is mighty to fight your battle. Jehovah Gibbor will fight for you—if you let Him. He needs your participation. So the question still stands:

What are you feeding?

Reflection Question:

What appetite have you been feeding that is weakening your spirit?

____

____

____

____

____

Challenge:

Ask the HOLY SPIRIT to expose the roots of your cravings. Don't just confess the fruit—get to the root. Then ask for wisdom on how to starve the appetite and strengthen your spirit.

PROPHETIC DECREE

Father,

I come to You. Just like Adam, I have been created in Your image—the image that gives me access to supernatural DNA. Through You I will run through troops and leap over walls. I have power over all the enemy—including habits and behaviors connected to destructive appetites.

I break agreement with every appetite that wars against my spirit. I renounce every craving, habit, trauma, and preference that's been feeding my flesh and starving my purpose. I declare: I will not be ruled by my desires—I will be ruled by the SPIRIT of the LIVING GOD.

I silence every lie that says I can't be free. I reject every identity shaped in addiction, indulgence, or emotional survival. I declare: I am not what I fed—I am what GOD formed. I was made in His image and created to glorify Him—not gratify my flesh.

LORD, help me want You more than I want relief. Help me desire You more than I desire comfort. I choose to feed my spirit with Your Word, Your presence, and Your truth. I will not live emaciated in the spirit. I will be strengthened, empowered, and aligned. I will not fall for false satisfaction.

I declare that freedom is available and I say yes. I say yes to hunger for righteousness. I say yes to transformation. I say yes to the process of being made whole. I am not a prisoner to appetite—I am a temple of the HOLY SPIRIT. I am healed and whole in JESUS CHRIST name, Amen.

This decree was influenced by:
Galatians 5:16, Matthew 5:6, and Romans 8:13.

Thoughts:

Thoughts:

Thoughts:

DAY TWENTY-THREE

1 Samuel 17:48 AMPC

When the Philistine came forward to meet David, David ran quickly toward the battle line to meet the Philistine.

Let's Talk About It:

What have you backed down from? What has seemed so overwhelming that you felt like David facing Goliath?

Well—take a page out of his book: Run toward it.

David didn't run from the battle—he ran into it. He didn't pause to assess how tall Goliath was or what others were saying. He had already spent time in the secret place, in the presence of the Most High. That presence became his courage. His confidence didn't come from the crowd—it came from his consecration.

The same is true for you. The more time you spend with GOD, the less you'll be shaken by giants. Goliath may have looked big, but David's GOD was bigger. That's the revelation that makes you run toward what you used to run from. Not because you feel strong—but because you know where your strength comes from.

Don't let the size of the battle intimidate you. Let the presence of the LORD activate you. Like David, you don't need man's armor—you just need GOD's anointing.

Reflection Question:

What fear or hesitation is keeping you from running toward the thing GOD has already equipped you to conquer?

Challenge:

Spend time this week sitting in the presence of GOD. Ask the HOLY SPIRIT to reveal the "giants" you've been avoiding—and then speak faith over them. Prepare to run.

PROPHETIC DECREE

Father, make me a person after Your heart. Let my desires long for Your presence. Transform me from the inside out. I want Your presence to be my portion in everything. Be my courage!

I renounce every fear that tried to paralyze my forward motion. I rebuke the voice of intimidation that tried to make me forget who I am. I break agreement with every generational lie that said I wasn't strong enough, smart enough, or worthy enough to fight. I silence the sound of defeat and step into the sound of victory.

I prophesy over myself: I will run. I will not delay. I will not hide. I will not retreat. I am aligned with divine timing and armed with divine authority. I run toward the very thing that tried to run me down. My aim is anointed. My footing is firm. And my GOD is faithful.
I thank You, LORD, for giving me eyes to see past the obstacle and ears to hear the strategy of Heaven. I thank You that giants are falling and victories are being secured. This is not just a battle—it's a setup for glory.

I am healed and whole in JESUS CHRIST name, Amen.

This prophetic decree was influenced by:
1 Samuel 17:48 AMPC

Thoughts:

Thoughts:

DAY TWENTY-FOUR

Philippians 4:8 KJV

Finally, brethren, whatsoever things are true, whatsoever things are honest, whatsoever things are just, whatsoever things are pure, whatsoever things are lovely, whatsoever things are of good report; if there be any virtue, and if there be any praise, think on these things.

Let's Talk About It:

Negativity isn't always a personality issue—it can be a protective mechanism. When life has taught you through repeated disappointment to expect the worst, your mind can develop what psychologists call negativity bias. This means you start focusing on what's wrong, what might go wrong, or what has gone wrong more than what is true, good, or hopeful.

But that's not how GOD designed your mind to function.

Paul's words in Philippians 4:8 are more than positive thinking—they're prophetic discipline. He's giving us a command to be intentional with our thoughts. Why? Because unchecked thoughts become silent agreements. And agreements shape expectations. If you keep expecting the worst, you'll find yourself bracing for heartbreak even when GOD is trying to bless you.

You are not crazy for struggling to think positive. Life has taught you to expect pain. But it's time to let the HOLY SPIRIT retrain your mind. Your brain is not broken. But it does need rewiring. And that happens when we stop rehearsing what hurt us and start declaring what heals us. The good report is not denial—it's deliverance.

This verse doesn't say we won't face bad news. It says: focus on what is lovely, pure, honest, praiseworthy, and true. That's not blind optimism—it's spiritual strategy.

Reflection Question:

What negative thought pattern has been your default, and how did life train you to expect that?

Challenge:

Ask the HOLY SPIRIT to highlight one expectation you've carried that was formed by trauma—not truth. Replace it today with a declaration rooted in Philippians 4:8.

____

____

____

____

____

PROPHETIC DECREE

Father, in the name of JESUS CHRIST, I bring You the battlefield of my mind. I surrender every thought that was shaped by trauma, not truth. I uproot every lie I rehearsed to protect myself. I break agreement with the negativity bias that taught me to expect disappointment before it even arrived. I renounce the strongholds built by fear, failure, and false narratives. I declare: the cycle is broken now.

I receive the mind of CHRIST. I receive divine discipline in my thinking. I declare my thoughts are being rewired with what is lovely, what is pure, what is of good report. I release the obsession with what could go wrong, and I embrace the possibility of what could go right. I will not brace for heartbreak—I will believe for healing. I will not partner with panic—I will meditate on praise.

I cast down every imagination that lifts itself against the knowledge of GOD. I reject the predictions of pain. I replace them with promises of peace. My mind is not a playground for the enemy—it is the sanctuary of the HOLY SPIRIT. I will think different. I will expect different. I will experience different.

This is not just a mindset shift—it is a supernatural exchange. I am healed and whole in JESUS CHRIST name, Amen.

This prophetic decree was influenced by:
Philippians 4:8 and 2 Corinthians 10:5

Thoughts:

Thoughts:

Thoughts:

DAY TWENTY-FIVE

Proverbs 27:7 AMPC

He who is satiated loathes honey, but to the hungry soul every bitter thing is sweet.

Let's Talk About It:

Hunger has a way of lying to you. When your soul is starving, even the bitter can start to look sweet. Proverbs 27:7 makes this painfully clear—"He who is satiated loathes honey, but to the hungry soul every bitter thing is sweet." In other words, when you're full, even the good stuff doesn't move you. But when you're starving—emotionally, spiritually, relationally—you'll start calling things sweet that are actually killing you.

That's what desperation does. It distorts discernment. Some of us have stayed in toxic relationships, returned to harmful habits, or ignored spiritual red flags—not because we didn't know better, but because we were hungry. Hungry for affirmation. Hungry for connection. Hungry for relief. And when you don't bring that hunger to GOD, the enemy will offer you something bitter—and lie to you until you swallow it.

But first, you've got to admit you're hungry. You've got to confront the cravings that you've tried to silence. You've got to stop pretending you're full just because you're functioning. GOD doesn't bless fake fullness—He blesses honest hunger.

The hunger is real, but misdirected. We are told, "Blessed are they which do hunger and thirst after righteousness: for they shall be filled." The only guarantee to be filled is connected to what we long for. Bring your appetite back to GOD. Let Him fill you—so you stop reaching for what was never meant to satisfy.

Reflection Question:

What have you been calling "sweet" that's really bitter? What craving have you been trying to satisfy outside of GOD?

Challenge:

Sit with the HOLY SPIRIT and ask Him to reveal any area where your hunger has been misdirected. Don't rush. Be honest. Then write down what healthy spiritual hunger looks like for you and ask the LORD to reset your appetite.

____

____

____

____

____

____

PROPHETIC DECREE

I confess that I've chased after things that couldn't fill me. I've swallowed what was bitter, thinking it was sweet. But today, I return to the Source. I bring my appetite back to GOD. I will not live on emotional junk food or settle for temporary fixes. I renounce every craving that wars against my purpose. I break agreement with soul ties that fed my dysfunction but starved my spirit.

I declare: I am not desperate—I am directed. I am not empty—I am being filled with righteousness, peace, and joy in the HOLY SPIRIT. I refuse to feast on bitterness. I will no longer call trauma love, or call survival peace. I will hunger for what heals. I will thirst for what restores. I am not tricked by the taste—I am anchored in truth.

Thank You, LORD, for resetting my appetite and restoring my soul. I am healed and whole in JESUS CHRIST name, Amen.

This prophetic decree was influenced by:
Proverbs 27:7 AMPC and Matthew 5:6

Thoughts:

Thoughts:

Thoughts:

DAY TWENTY-SIX

Proverbs 24:33–34 AMPC

Yet a little sleep, a little slumber, a little folding of the hands to lie down and sleep—So will your poverty come like a robber or one who travels [with slowly but surely approaching steps] and your want like an armed man [making you helpless].

Let's Talk About It:

Sleep can be holy—or harmful. It all depends on what it's hiding.

Proverbs 24:33–34 warns us that "a little sleep... a little folding of the hands" can open the door to poverty. But this isn't just talking about laziness—it's exposing a pattern of avoidance. For some, sleep is no longer about rest—it's a mechanism to escape pain. An emotional sedative. When life feels too heavy, sleep can become a hiding place. But when that pattern becomes our norm, poverty—emotional, spiritual, or financial—creeps in like a thief.

This isn't just about naps. It's about numbness. Some people oversleep because they're overwhelmed. Sleep becomes a safe place to disconnect from what they don't know how to fix. But the danger is this: sleep that was meant to restore can start to restrain. It becomes a cycle of delay, and delay becomes defeat.

Yet we also see another truth in Scripture. When Elijah was overwhelmed, depressed, and ready to die, GOD didn't rebuke him—He gave him a nap and a snack. That moment shows us something powerful: balance is the key. Sleep in right alignment is a reset. It can refresh you, refocus you, and fuel you for your next assignment. But when sleep becomes a substitute for surrender or a replacement for responsibility, it opens the door to loss.

This is a wake-up call—literally and spiritually. Your purpose requires alertness. Your healing requires presence. You don't get whole by sleeping through the process. But you also don't get whole by grinding without grace. The answer isn't no sleep—it's aligned sleep. It's resting in HIM, not running from life.

Reflection Question:

Are there moments where sleep has become your escape instead of your reset?

Challenge:

Ask the HOLY SPIRIT to show you where your rest has turned into resistance. Invite Him to help you reset your rhythm and reintroduce alignment into your sleep, your habits, and your healing.

____

____

____

____

____

____

PROPHETIC DECREE

Father, I surrender my schedule, my avoidance, and my escape mechanisms. I break agreement with cycles of delay that have disguised themselves as rest. I will not partner with apathy or the lies that tell me hiding is healing. I release the false peace of avoidance and receive the real peace of Your presence.

I declare: I am not stuck. I am not asleep in my purpose. I shake off the slumber. I renounce spiritual drowsiness. I rise in alertness, wisdom, and discernment. I will not fold my hands in emotional defeat—I will lift them in prophetic alignment. I declare that balance will lead me, discipline will protect me, and surrender will anchor me.

Thank You, LORD, that You give Your beloved sleep. But You also call us to rise. I receive Your grace to rest with purpose and awaken with power. I am healed and whole in JESUS CHRIST name, Amen.

This prophetic decree was influenced by:
Proverbs 24:33–34 AMPC

Thoughts:

Thoughts:

Thoughts:

DAY TWENTY-SEVEN

1 Samuel 17:38–39 KJV

And Saul armed David with his armour, and he put an helmet of brass upon his head; also he armed him with a coat of mail. And David girded his sword upon his armour, and he assayed to go; for he had not proved it. And David said unto Saul, I cannot go with these; for I have not proved them. And David put them off him.

Let's Talk About It:

Have you ever had someone try to help you in a way that only made things worse? Not because they were malicious—but because they didn't understand your purpose? That's what happened here. People weren't trying to sabotage you. They were trying to protect you. But their method wasn't made for your mission. They offered what they would use. They suggested what they needed to feel safe. Their advice came from experience—but not necessarily from revelation. They were trying to help you fight, but ended up loading you down with their fear, their strategy, their armor.

That's a heavy weight to carry—especially when it's wrapped in good intentions. But here's the tension: just because something works for someone else doesn't mean it's what GOD is calling you to carry.

David didn't dishonor Saul. He didn't rebel. He simply recognized that the armor didn't fit. Not every no is disobedience. Sometimes it's discernment. But here's the balance: discernment isn't an excuse for isolation.

We're not called to be rogue or reckless. The Bible is clear: in the multitude of counsel there is safety. We need mentors. We need voices of wisdom. But we also need the courage to admit when something—no matter how respected—hasn't been proven in our process. David didn't win battles with swords. He won them with slings. He didn't reject counsel—he just trusted what had already been tested with GOD.

This is the place of maturity: where you can honor the people who poured into you while still obeying the path GOD marked out for you. Where you can value the voices around you without being hijacked by their fears. Your purpose comes with tools—and they're not always what others expect. The oil on your life will always recognize what fits. Trust it.

And remember: GOD will always confirm His way with victory.

Reflection Question:

Where in your life have you been pressured to carry something that doesn't fit—just to make others comfortable?

Challenge:

Ask HOLY SPIRIT to show you what has been proven in your journey and what needs to be laid down. Journal what fits your anointing—not your audience.

PROPHETIC DECREE

I refuse to carry what doesn't fit. I will not wear what was never proven in my process. I break agreement with pressure that comes from people's fears and projections. I renounce every strategy that came from flesh and not from GOD. I declare that I have the courage to walk in what has been tested in my history with GOD.

I honor the voices of wisdom You've placed around me, LORD—but I will not confuse tradition with instruction. I will not forfeit victory trying to fight battles with someone else's armor. I trust the weapons You've trained me with. I choose function over familiarity. I pick up what You've placed in my hands. I drop what never belonged to me.

Thank You, LORD, for confirming Your way with results. Thank You for aligning me with divine strategy. Thank You for reminding me that obedience is my protection and purpose is my provision. I walk boldly into battle with what You've tested in me. I am healed and whole in JESUS CHRIST name, Amen.

This prophetic decree was influenced by:
1 Samuel 17:38–39

Thoughts:

Thoughts:

Thoughts:

DAY TWENTY-EIGHT

Philippians 4:13 AMPC

I have strength for all things in CHRIST Who empowers me [I am ready for anything and equal to anything through Him Who infuses inner strength into me; I am self-sufficient in CHRIST's sufficiency].

Let's Talk About It:

Low self-esteem doesn't need a demon—it just needs a narrative. And for many of us, that narrative was formed early. We were taught that we weren't enough. That we didn't have what it takes. That other people were better equipped. And over time, we internalized that belief and began to act like it was true. But this verse reminds us of a radical truth: our sufficiency is not of ourselves—it is of GOD.

The enemy doesn't have to launch a full attack—he just has to convince you to doubt your own worth. If he can get you to disqualify yourself, you'll never step into what GOD already said is yours. You'll back down from opportunities. You'll second-guess divine instruction. You'll compare, shrink, and sabotage. And he won't have to do a thing—because you'll do it to yourself.

But not anymore.

You were never meant to do it in your own strength. You were never meant to be enough without CHRIST. The power was never supposed to come from you—it flows through you. That's why the Word doesn't say "I can do all things because I'm gifted." It says "through Him who infuses inner strength into me." That's the shift. You don't work your way into confidence—you surrender into sufficiency.

Not only are you able—you are equipped. You are infused with inner strength. So next time the enemy whispers that you're not ready, remind him what the Word says: I am ready for anything and equal to anything through CHRIST.

Let that become your declaration. Ready? Let's go.

Reflection Question:

What lie about your inadequacy have you believed, and how has it held you back from pursuing GOD's purpose for your life?

Challenge:

Declare Philippians 4:13 over yourself throughout the day. Every time you feel fear, hesitation, or unworthiness rising up—speak the truth of GOD's Word out loud and let it reframe your confidence.

PROPHETIC DECREE

FATHER, I confess that I have believed the lie that I'm not enough. I've measured myself by my own strength instead of Yours. I've shrunk when You called me to stand tall. But today, I align with Your Word. I declare that I am ready for anything and equal to anything through JESUS CHRIST who infuses me with strength.

I break agreement with every voice that told me I couldn't. I renounce the mindset of fear, failure, and inadequacy. I silence the enemy's accusations and my own inner critic. I receive the truth: I am not underqualified—I am GOD-chosen. I am not insufficient—I am CHRIST-equipped.

I step into divine boldness. I move with holy confidence. I embrace supernatural capacity. Not because of who I am—but because of who You are in me. I declare: I will no longer disqualify myself from destiny. I am empowered. I am enough in You. I am healed and whole in JESUS CHRIST name, Amen.

This prophetic decree was influenced by:
Philippians 4:13 AMPC.

Thoughts:

Thoughts:

Thoughts:

DAY TWENTY-NINE

Acts 7:59–60 AMPC

And while they were stoning Stephen, he prayed, "LORD JESUS, receive and accept and welcome my spirit!" And falling on his knees, he cried out loudly, "LORD, fix not this sin upon them!" And when he had said this, he fell asleep [in death].

Let's Talk About It:

What's your first instinct when someone attacks you—defend yourself, curse them, or pray they get what's coming?

We've all felt the sting of betrayal, the weight of injustice, or the ache of unfair treatment. And it's natural to want GOD to step in with judgment. But Stephen—full of the HOLY SPIRIT—shows us a different response: mercy. Grace. Forgiveness.

He didn't call down fire. He asked for forgiveness for the very people killing him. That wasn't weakness—it was evidence of divine strength. The more filled you are with the SPIRIT, the more your instincts shift. You stop wanting revenge. You start wanting redemption.

This doesn't mean you ignore injustice or pretend it doesn't hurt. It means you refuse to be mastered by offense. You refuse to let the attack change your identity. Mercy is not the absence of pain—it's the overflow of presence. And when you're full of the HOLY SPIRIT, mercy wins.

Reflection Question:

What offense have you been carrying that reveals you've been operating from pain instead of presence?

____

____

____

____

____

Challenge

Ask the HOLY SPIRIT to show you where your responses have been based on emotion instead of empowerment. Invite Him to help you release the need for retaliation and receive His grace to respond like JESUS.

____

____

____

PROPHETIC DECREE

FATHER, I repent for every time I wanted revenge more than restoration. I break agreement with bitterness, retaliation, and resentment. I refuse to partner with pain as my prophet. I will not let the actions of others determine my identity or dictate my response.

I declare: I am full of the HOLY SPIRIT. I do not respond from flesh—I respond from faith. I speak grace instead of cursing. I choose forgiveness over fixation. I declare that mercy flows out of me because the SPIRIT lives within me. I silence the voice that says people deserve my punishment and amplify the voice that says they need Your presence.

Even in the face of attack, I choose love. Even in the presence of betrayal, I choose freedom. I release every offender into Your hands, and I receive the peace that comes from obedience.

I am healed and whole in JESUS CHRIST name, Amen.

This prophetic decree was influenced by:
Acts 7:59–60 AMPC, Ephesians 6:12, and Matthew 5:7

Thoughts:

Thoughts:

DAY THIRTY

Proverbs 6:27 AMPC

Can a man take fire in his bosom and his clothes not be burned?

Let's Talk About It:

Can a man take fire in his bosom and not be burned?

That's not just a rhetorical question. It's a warning. Proverbs 6:27 isn't simply about adultery—it's about proximity to danger. It reminds us that some things, no matter how justified they feel, will burn us if we hold them too closely. And today, we're focusing on one of those things: anger.

Anger in itself is not a sin. Even JESUS was angry. The problem isn't the feeling—it's how we handle it. When anger becomes our counselor, our motivator, or our defense mechanism, we're holding fire to our chest and expecting not to get burned.

Some of us have normalized anger as fuel. We say, "It's what keeps me going." But when anger is your energy, it eventually becomes your enemy. It burns bridges, blurs boundaries, and blinds us to the presence of GOD. You can't cling to HIS peace and your rage at the same time.

This doesn't mean you have to pretend you're not mad. It means you must learn to process your anger with the HOLY SPIRIT, not apart from Him. HIS fire purifies—ours consumes. HIS fire restores—ours destroys.

So if you've been holding anger close, ask yourself: what has it cost me? Has it burned through your joy? Has it scorched your relationships? Has it singed your purpose? Then take that fire, and put it on the altar. Let GOD make it holy. Let Him turn it into passion for healing, not destruction.

Reflection Question:

What anger have you been carrying that is starting to burn holes in your peace, your relationships, or your purpose?

____

____

____

____

____

Challenge:

Ask the HOLY SPIRIT to reveal any place in your life where anger has been smoldering beneath the surface. Invite Him to put out the flame and teach you how to respond from righteousness—not reaction.

PROPHETIC DECREE

Father, I bring You every ember of anger I've been carrying—hidden, justified, or ignored. I repent for letting rage dwell where only Your presence belongs. I break agreement with resentment, retaliation, and the belief that my anger makes me strong. I declare: I will not burn for bitterness. I will burn for purpose.

I lay down the fire that consumes and pick up the fire that purifies. I renounce the lie that I have to stay angry to stay powerful. I cast off the torment that anger disguised as protection. I will not nurse my grudges—I will nurture Your glory. I break agreement with pettiness that keeps offense alive and fuels slow-burning fury. I declare that I am not ruled by emotion but governed by Your SPIRIT.

The same GOD who made fire walk with the Hebrew boys walks with me now. I will not let anger scorch my peace, sear my joy, or stain my future. The fire on Your altar will be my only flame. I am healed and whole in JESUS CHRIST name, Amen.

This prophetic decree was influenced by:
Proverbs 6:27, Ephesians 4:26–27, and James 1:20

Thoughts:

Thoughts:

Thoughts:

DAY THIRTY-ONE

Jeremiah 17:14 AMPC

Heal me, O LORD, and I shall be healed; save me, and I shall be saved, for You are my praise.

Let's Talk About It:

When GOD does it—it's done.

Jeremiah's cry in this verse is both a declaration and a surrender. He doesn't say, "I'll try to be healed" or "I'll see if it works." He says, "Heal me… and I shall be healed." That's the kind of healing that comes when GOD is your Source. There's no residue of brokenness, no remaining infection, no half-done work. When GOD heals, it is complete.

But here's the truth: healing requires participation. It's not about striving, but surrendering. You can't say "heal me" while still holding on to what broke you. You can't ask to be made whole while clinging to fractured thinking. Healing is a divine work—but it's also a human decision.

This 31-day journey has been about more than information. It's been about transformation. But don't stop here. This isn't the end of your healing—it's the launching pad for wholeness. The deeper you go into His presence, the higher He'll take you in purpose.

You've touched the hem, but there's more. You've cried out like Jeremiah, now keep walking like the woman with the issue of blood—pressing until nothing is left but wholeness.

Don't settle for "better." Press for "complete." And don't do it alone—invite the HOLY SPIRIT to lead the rest of the journey. You've seen what happens when you participate in the process. Now keep going.

Reflection Question:

Where do you still need to give GOD permission to heal you completely—and what are you willing to release so that healing can happen?

____

____

____

____

____

Challenge

Declare to the LORD today that you are not finished healing and that you're ready to keep participating. Ask the HOLY SPIRIT for your next step toward becoming whole—body, soul, and spirit.

PROPHETIC DECREE

Father, I thank You for beginning the healing process in me—but I will not stop here. I surrender every wound, every scar, and every survival mechanism I've used to protect myself. I declare: You are the GOD who finishes what You start. You are the GOD who heals completely.

I break agreement with partial healing, emotional numbness, and spiritual avoidance. I renounce the lie that this is as good as it gets. I say: no more cycles, no more stuck seasons, no more settling for symptom relief. I want the root healed. I want the pain replaced with purpose. I want You to have full access to every room in my soul.

I give You permission to continue. I declare: this is not the end—it's the elevation. This is not the finish line—it's the beginning of a new flow. The GOD who heals me will walk with me, lead me, and mature me until nothing in me looks like where I came from.

I am healed and whole in JESUS CHRIST name, Amen.

This prophetic decree was influenced by:
Jeremiah 17:14 AMPC, Philippians 1:6, and Luke 8:48

Thoughts:

Thoughts:

Thoughts:

Just The Beginning

Father GOD,

Thank You for beginning a good work in me. Thank You for not leaving me stuck in the wounds of my past or the patterns of pain that shaped me. Thank You for using this journey not to finish the healing—but to start it. This is the beginning of becoming whole. This is the opening of a new chapter. And I do not take it lightly.

I confess that healing is not always instant—but it is always available. And I choose to keep walking. I choose to keep surrendering. I choose to let Your truth shape my behavior, rewire my responses, and change my decisions. I don't just want to feel better —I want to live different.

So I seal what You have awakened. I wrap it in obedience. I surround it with wisdom. I anchor it in truth. I declare that no lie, no trigger, no old pattern will uproot what You planted in these pages. I may still have questions, but I have enough clarity to move. I may still have scars, but they are proof I survived. I may still have healing to do, but I say YES to the process.

Teach me to walk out what You revealed. Help me not just cry at altars but build a life of alignment. Let my language shift. Let my habits shift. Let my relationships shift. Let my appetite shift. Let my atmosphere shift. Make my life an echo of the healing You've begun.

I will not romanticize dysfunction. I will not make excuses for what You've given me authority to overcome. I will not be casual with what cost You everything. I will continue. I will apply. I will obey.

I am not just healed—I am being made whole. And wholeness has requirements. I accept them.

Seal this work, HOLY SPIRIT. Protect it. Grow it. And lead me into all truth.

In JESUS CHRIST name, Amen.

We want to walk with you.

Healing was never meant to be a solo journey. If you've been stirred, challenged, or transformed during these 31 days, don't stop now. Join our Healed & Whole community on Telegram—a safe, Spirit-filled space where we keep growing together, pray together, and press toward complete healing and wholeness. You don't have to do this alone.

Telegram Group: **Healed & Whole**
Link: https://t.me/+SCNaXelEDZc5ZWMx
Or Scan this QR:

About the Author

Apostle Idella Liselle is a prophetic voice raised for such a time as this—called to misfits, wounded warriors, and those who love GOD but still carry scars. With over 35 years in ministry, 30 years in the prophetic, and 14 years of coaching, she is a walking testimony of what it means to be healed and whole for real.

Once silenced by trauma and boxed in by religion, Apostle Idella now leads with boldness and compassion, reminding others that healing is not a side-note to purpose—it's the pathway to it. As the founder of REAL Ministries, she leads a global online movement committed to creating safe spaces for inner healing, deliverance, and legacy-level transformation.

Through platforms like Called & Confident Collective, The Be Academy, and her signature Hearing GOD's Voice trainings, she equips people to confront emotional cycles, reclaim identity, and walk in supernatural wholeness. Whether she's preaching, mentoring, or coaching, her voice carries the weight of someone who's lived what she teaches.

Apostle Idella is the CEO of Idella Liselle CEO, a powerhouse brand where she mentors leaders, coaches visionaries, builds systems, develops Kingdom-minded entrepreneurs, and provides publishing strategies through her publishing company, PublishHer, to help others launch movements, businesses, and books that carry lasting impact.

She now resides in the Dominican Republic with her husband, Pastor Nidio Familia, serving on the frontlines of Kingdom work—building a center called The Place of Solutions, with a vision to restore and establish hope within the community by providing daycare that offsets costs for parents, afterschool programs that empower youth, and business development initiatives that equip individuals to rise in purpose.

Her life is a living decree: Healing is possible. Wholeness is real. And the journey is worth it.

If this devotional stirred something in you—good. That means healing has begun. You're not just a reader—you're one of us. A seeker. A fighter. A finisher. The kind of person who doesn't want surface religion or emotional hype—but deep, lasting transformation.

At REAL Ministries, we are cultivating a global community of people being **Healed and Whole 4REAL.** We are an online movement and a safe place for those who've been dropped, silenced, overlooked, or burned out by religion. This is for those who love GOD but still carry wounds. Who are ready to confront what hurt them, heal from what broke them, and become whole from the inside out.

If you're looking for a place where the HOLY SPIRIT is welcome, where truth is taught with authority, and where healing is not just possible but expected—then welcome home.

This is your invitation to walk it out with a tribe that gets it. A tribe that isn't pretending. A tribe that is healing on purpose.

Join the movement. Be part of something that doesn't just speak to pain—but calls you into purpose.

Visit us at **www.REALMinistries.Online**
Text **678-712-8942** to join our text community.
You don't have to do this alone. There's a tribe for you.

We are REAL.
We are healing.
We are becoming whole.
We are Healed and Whole 4REAL.

Work With Me: Healed and Whole Connection

Ready to own you?

Ready to be authentically the YOU GOD created — or share the unfiltered message He gave you? It's time to stop playing small. Your voice, your message, your product, and your service deserve to be seen, heard, and received by the people who need them most. I help you find your superpower, remove what's been in the way, and set you in position to walk out exactly what GOD has called you to do and become.

If *Healed and Whole* was your wake-up call, this journey is your next step. The same boldness it takes to claim your healing is the boldness it takes to release your voice, your story, and your GOD-given assignment to the world. Let's partner to remove the obstacles and put you in position to be and do exactly what GOD has ordained.

Publishing: For Individuals and Compilations

GOD told you to write it — what are you waiting for? Delay is still disobedience, whether it's caused by fear or simply not knowing how to get it done. We partner with authors who carry a message from GOD and do the heavy lifting to bring it to life. Whether you need someone to build or refine your story, or you're ready to move forward but just don't know the next step, we can help.

Just getting started? A compilation project may be the perfect first step. You'll contribute one powerful chapter and still gain the credibility and reach of a published author. Compilations allow you to connect with new audiences, open doors for ministry, business, and speaking opportunities — without the time and pressure of writing an entire book. Many authors use their compilation chapter as a launching pad for future solo projects and as a tool to establish their voice in the marketplace.

If you've walked through transformation and carry wisdom that could set someone else free, your story is more than testimony — it's a tool. Publishing with us allows you to turn that hard-won journey into something that ministers long after you've spoken it, reaching hearts and lives you may never meet in person.

Everything we do echoes Kingdom — from the cover design to the process itself. Let us help you finish what GOD started in you and complete your obedience. Someone is waiting for your yes.

Coaching: Life, Ministry, and Business.

Life Coaching

Life doesn't change until you do. In life coaching, we uncover your "why" and map out your "how," dismantling defaults that keep you stuck and aligning every move with your GOD-given purpose. You'll gain the clarity and courage to live with intentionality, not survival.

Business Coaching

Your business should carry your vision — and your vision should produce results. In business coaching, we identify products and services that speak to your audience, ensure your brand is consistent across every touchpoint, and create strategies that generate both influence and income.

If your journey has taken you through deep places of healing and breakthrough, you already know how powerful change can be. Coaching with us helps you take that same transformation and apply it to your calling — so you're not just surviving your story, you're leading from it.

No matter the focus, my coaching equips you to ask the right questions, make the right moves, and achieve results that last.

Let's make it happen. Your obedience could be someone else's breakthrough.

Contact us today:

770-727-1669

www.idellaliselle.com

BE Academy: 3-Day Intensive Workshop

Three days. One mission. Your breakthrough.

If *Healed and Whole* helped you confront pain, trauma, and the wounds beneath the surface, this intensive is your next step. The BE Academy Intensive takes that same healing journey and pushes it further — stripping away the lies, cycles, and sabotage that have been quietly robbing you of your destiny, and replacing them with unshakable identity, strategy, and tangible action.

Day 1 – Take the Inner Me Out
Go deeper into the excavation process you started in Healed and Whole. Confront the dysfunction, behavior patterns, and generational cycles that have been holding you hostage, and break their grip for good.

Day 2 – LOUD on Purpose
If healing made you whole, this is where you become bold. No more shrinking. No more apologizing for who you are. Reclaim your GOD-given identity, own your call, and turn up the volume on the life you were designed to live.

Day 3 – Plan, Prepare, Produce
Healing without action leaves you in the same place. We don't just talk about change — we produce it. You'll leave with a step-by-step strategy and real-world tools to turn what you've uncovered here into results you can see, touch, and measure.

Along the way, you'll join in a final-day service project that allows you to pour back into the local community — because true wholeness is never selfish; it's about the Kingdom impact you carry.

An Experience Like No Other

These intensives are intentionally limited in size to ensure you get deep, personal attention and breakthrough. They include your accommodations, meals, training materials, transportation, and service project participation — so you can focus fully on your transformation.

For upcoming dates or to host a BE Academy Experience in your city, visit www.idellaliselle.com.

Hearing GOD's Voice: 5-Week Intensive Training

Since 2010, *Hearing GOD's Voice* has become a catalytic experience for those ready to stop wondering if GOD is speaking—and start walking in the confidence that He is.

This 5-week intensive doesn't just teach—it activates. Rooted in biblical foundation and prophetic clarity, this course is designed to strip away the noise, doubt, and default systems that muffle GOD's voice. Fear, insecurity, and religious performance are confronted head-on—because real clarity comes when the clutter is cleared.

Participants are guided through scriptural anchors that reveal how GOD speaks, how He has always spoken, and how to discern His voice through identity, not emotion. You'll learn to recognize the tone, rhythm, and nature of GOD's voice—while confronting every internal lie that told you it was just your imagination.

What sets this class apart are the Spirit-led activations that stretch spiritual sensitivity and eliminate reliance on human cues. These exercises cultivate a bold, pure dependence on HOLY SPIRIT—and challenge each participant to trust what they hear before they see. You are pulled to hear GOD beyond what you know. We remove man's crutches to open your spiritual ears. These activations don't just invite you forward—they push you off the deep end. And while the world says your only options are to sink or swim, through these classes the SPIRIT reveals a third, better option: walk on water.

Even those who didn't finish the course have testified: *"Something shifted. I hear GOD more clearly now."*

If you've ever second-guessed whether it was GOD or your own thoughts—this class is your next step.

To register for the next Hearing GOD's Voice training, visit <u>www.idellaliselle.com</u> for more information.

Other Books by Idella Liselle

Crazy on Purpose

Being "normal" was never the assignment. In this bold and prophetic work, Idella Liselle calls out the systems, mindsets, and fears that keep you playing small — and equips you to step fully into the GOD-designed life you were born to live. Packed with raw truth, biblical insight, and unapologetic challenges, this book will dismantle excuses and ignite action. If you are ready to stop blending in, break out of the box, and walk in your Kingdom calling without apology, this book will show you how.

The Rantings of a Crazy Woman: Words of Life and Transformation

Unfiltered, unpolished, and undeniably prophetic — this book is Idella Liselle's unapologetic call to break cycles, confront comfort zones, and disrupt anything that keeps you from your GOD-ordained destiny. With raw stories, sharp truths, and SPIRIT-led insight, she challenges the reader to stop playing safe and start living bold. If you are ready to embrace your call, confront your excuses, and live with unapologetic Kingdom audacity, this book will push you there.

Is Church Killing You?

A bold and unflinching look at the deception and compromise that breed church hurt. Drawing from personal encounters and prophetic insight, Idella Liselle exposes the lies that keep believers bound, challenges the culture of excusing sin, and calls the Body of CHRIST back to truth, accountability, and healing. This book doesn't just confront the problem — it redefines what ministry should be. If you are seeking freedom from church wounds and a fresh, GOD-centered vision for ministry, this book will help lead you there.

Thoughts of a Virgin

A raw, unapologetic look at the dirty little secrets the church doesn't want to talk about. Idella Liselle shares her personal journey from early sexual exposure to being delivered from masturbation and pornography, and the fight to walk in true freedom. She dismantles shame with honesty and points to the GOD who heals the hidden places. If you are ready to confront private battles and break free from the chains of sexual sin, this book will show you the way to freedom.

The Diary of a BAD Girl, Kingdom Style

Proof that GOD can turn the mess into a mandate. In this unapologetic compilation, Idella Liselle and other women share raw testimonies of early exposure to perversion, spiritual battles, and the traps that were meant to destroy them — and how GOD used it all for good. These are stories of warfare, deliverance, and undeniable transformation. If you've ever wondered whether your past disqualifies you, this book will show you how GOD redeems, restores, and reassigns.

Wife Material

Just as we are called to be clothed with CHRIST, we are also called to be clothed as wives — in character, strength, and Kingdom identity. In this powerful compilation, wives share real stories of how GOD met them in their marriages, through trials, victories, and unexpected lessons. These testimonies reveal that being "wife material" is less about perfection and more about divine preparation. If you desire a marriage shaped by GOD's presence and strengthened by His wisdom, this book will inspire and equip you.

Legacy: The Prayers of a Mother

This is not a keepsake — it's a strategy. In this powerful book, Idella Liselle delivers full, targeted prayers birthed from a legacy of relentless intercession. Designed to cover children with precision and power, these prayers carry the kind of warfare that shifts atmospheres, breaks demonic assignments, and calls forth destiny. Woven with testimonies of supernatural answers to prayer, this book becomes a spiritual covering you can return to again and again. If you are ready to pray with power, precision, and results, this book will put the weapon in your hands.

Leverage: A Prophetic Planner

More than a planner — this is a prophetic strategy tool. Designed to help you ask the right questions for any prophetic word, assignment, or project, Leverage walks you through the process of turning revelation into results. With built-in guidance for discernment, timing, and execution, Idella Liselle equips you to move with clarity and precision. If you are ready to stop sitting on what GOD has spoken and start building it with strategy, this planner will show you how.

When people are restored, whole, and walking in their GOD-given purpose, they become unstoppable forces for change. That's the heart behind everything we do—whether it's giving widowed pastors' wives a place of dignity and safety at Grace House or mobilizing teams to serve and bring transformation through missions. Every hand that serves, every brick laid, every prayer prayed moves someone closer to the wholeness GOD intended. And you can be part of that story.

Grace House: Continuing a Legacy

Grace House was my mother's last assignment—a vision birthed from her heart as a pastor's wife who saw firsthand the lack of care given to pastors' widows after their husbands passed. She dreamed of building a place where these women could find safety, dignity, and rest when they had nowhere else to go. Grace House is not yet complete, but you can help finish what she started. You can send a construction team, contribute building materials, or donate toward its completion. Your support ensures this legacy lives on and serves the women who have served others so faithfully.

Scan to give to the Grace House Project

REAL Missions Global: Dominican Republic, Mexico, and Uganda

We believe missions aren't just trips—they're opportunities to transform lives, including your own. Our missions range from 3-day to 7-day experiences, and we can customize the focus based on your ministry's purpose. We work with small groups of 10 or fewer to ensure deep connection, flexibility, and meaningful impact. Whether it's serving children, empowering women, or advancing Kingdom projects, you'll leave marked by the work you've done and the people you've served. Join us in the Dominican Republic, Mexico, or Uganda, and become part of a mission that changes communities—and changes you.

Connect with us at www.realmissionsglobal.com

Be Part of the Mission: Become a Monthly Partner

Not everyone can join a missions or construction team in the Dominican Republic—but you can still make a life-changing impact. Your monthly donation goes toward the maintenance of what's already in place and the establishment of our future goals: a daycare to give children a safe and nurturing start, a medical clinic to meet urgent health needs, a Pastor Leadership Lab for mentoring and discipleship, women's business empowerment training, afterschool programs for local children, and more.

Just $60 a month covers the full daycare expense for a child once the program is operational, while $12 a month helps maintain our center so we can keep building toward these goals. Every gift moves us closer to a fully equipped community hub that will serve families, leaders, and future generations. Whether you're helping us build it or helping us sustain it, your partnership makes you a vital part of what GOD is doing through REAL Solutions.

Text to Give: 844-578-2358
Zelle: realministriesga@gmail.com
Give Online: www.REALMissionsglobal.com

Stay Connected with Idella Liselle

Transformation isn't built in one moment — it's cultivated through daily decisions, declarations, and divine strategies. Here's how you can stay directly connected, equipped, and activated:

Daily Prophetic Decrees

Text *DECREE* to *678-712-8942*

Receive daily prophetic decrees to speak over your life, ignite your faith, and align your atmosphere with GOD's truth.

Strategic Updates & Insights

Text *IDELLA* to *770-727-1669*

Get strategic updates, prophetic insights, and behind-the-scenes access to events, trainings, and Kingdom initiatives.

This is your personal direct line to the words, wisdom, and strategies that will keep you moving in purpose.

Sign up today — stay equipped for tomorrow.

Email

For speaking requests, coaching inquiries, publishing consultations, or general questions:

idellaliselle@gmail.com

Online

- Website: *www.idellaliselle.com*
- Social Media: Connect on Facebook, Instagram, TikTok, Linked In, and YouTube @IdellaLiselle

Booking

To schedule a Vision Discovery Call, visit:

https://idellaliselleceo.as.me/VisionDiscoveryCall

Made in the USA
Columbia, SC
19 January 2026

77013825R00122